A Practical Guide to Clinical Bacteriology

A Practical Guide to Clinical Bacteriology

J. R. Pattison

R. N. Gruneberg

J. Holton

G. L. Ridgway

G. Scott

A. P. R. Wilson

University College London Medical School and
University College London Hospitals, London, UK

JOHN WILEY & SONS
Chichester · New York · Brisbane · Toronto · Singapore

Other Wiley Editorial Offices

John Wiley & Sons, Inc., 605 Third Avenue,
New York, NY 10158-0012, USA

Jacaranda Wiley Ltd, 33 Park Road, Milton,
Queensland 4064, Australia

John Wiley & Sons (Canada) Ltd, 22 Worcester Road,
Rexdale, Ontario M9W 1L1, Canada

John Wiley & Sons (SEA) Pte Ltd, 37 Jalan Pemimpin #05-04,
Block B, Union Industrial Building, Singapore 2057

**This book is based on
Handbok i KliniskVirologi edited by Gunnar
Haukenes and Lars R Haaheim, 1983,
Universitetsforlaget, Norway.**

Library of Congress Cataloging-in-Publication Data
A practical guide to clinical bacteriology / edited by J. R. Pattison
    . . . [et al.].
        p.   cm.
    Includes index.
    ISBN 0–471–95288–5 (pbk.)
    1. Medical bacteriology.   I. Pattison, J. R. (John Ridley)
    [DNLM: 1. Bacteriology – handbooks.   QW 39 P8945 1995]
    QR46.P83   1995
    616'.014 – dc20
    DNLM/DLC
    for Library of Congress   94–39451
                                                      CIP

British Library Cataloguing in Publication Data

A catalogue record for this book is available from the British Library

ISBN 0 471 95288 5

Typeset in 10/12 Garamond by Vision Typesetting, Manchester
Printed and bound in Great Britain by Redwood Books Ltd, Trowbridge, Wiltshire

Contents

List of Contributors

Professor J. R. Pattison
Professor of Medical Microbiology, UCL Medical School, Rayne Institute, University Street, London WC1E 6JJ
Chapters 18, 28, 32, 36

Dr R. N. Gruneberg
Consultant Microbiologist, UCL Hospitals, Grafton Way, London WC1E 6AU
Chapters 3, 10, 11, 23, 29

Dr J. Holton
Senior Lecturer, UCL Medical School, Riding House Street, London W1
Chapters 1, 6, 13, 19, 21, 25, 31, 33

Dr G. L. Ridgway
Consultant Microbiologist, UCL Hospitals, Grafton Way, London, WC1E 6AU
Chapters 5, 7, 14, 22, 34, 35

Dr G. Scott
Consultant Microbiologist, UCL Hospitals, Grafton Way, London WC1E 6AU
Chapters 2, 8, 15, 17, 30

Dr A. P. R. Wilson
Senior Lecturer, UCL Medical School, Grafton Way, London WC1E 6AU
Chapters 4, 9, 12, 16, 20, 24, 26, 27

Preface

Bacteriology is an important core subject for students of medicine and the professions allied to medicine. Moreover the subject continues to evolve in relation to the discovery of new micro-organisms, new bacterial diseases, new antibiotics and new approaches to the control of bacterial infection. Thus there is a continual need amongst students and professionals in specialties other than medical microbiology and infectious diseases for concise readable accounts of the subject since the definitive texts often run to more than one volume. We were encouraged to produce such a concise account by the involvement of one of us in the sister book *A Practical Guide to Clinical Virology* and by the belief that no existing text is universally successful.

As far as possible we have worked to include only those aspects of the subject which we believe are necessary for a broad understanding of the subject. We begin with general chapters and emphasise the principles underlying the subject. In many ways these are the most important chapters since, if the general principles are understood, the changing detail for the future will be easily assimilated. However at any point in time there is a need to understand some of the detail of systematic bacteriology and the later chapters are intended to provide a succinct, readable and easily used account of the bacteria of medical importance. Each chapter has identical sections arranged in the same order so that the reader, once familiar with the book, will find it easy to use. There is a summary page at the beginning of each of the systematic bacteriology chapters for ease of revision of the topics. There is also a cartoon within each chapter which emphasises one aspect of importance in each chapter. We trust that these will both amuse and inform.

The book is intended for any student concerned with human disease but particularly medical students. We also believe it will be of use to qualified doctors and other professionals whenever they are stimulated by examinations or by clinical practice to revise their knowledge of the subject.

J. R. Pattison
R. N. Gruneberg
J. Holton
G. L. Ridgway
G. Scott
A. P. R. Wilson

London 1995

Acknowledgements

The authors gratefully acknowledge the skill of the artist Guy Venables who produced the cartoons, and the help they received from all their secretaries, especially Mrs Nanette Dyke.

1 Classification of Bacteria

In the classification of plants and animals the concept of a *species* is important for defining a group, reproductively isolated from other species, yet evolutionarily related to them. This is represented by a taxonomic hierarchy of Genus, Family, Order, etc. Bacteria replicate vegetatively, have few morphological characteristics and may share a common gene pool by the exchange of transposable genes. Thus, the evolutionary relationships between bacteria are not as clear as they are between other organisms, and the meaning of the term species in bacterial classification is a grouping based upon shared properties and not evolutionary relatedness.

Bacteria are *prokaryotic* micro-organisms. Algae, protozoa and fungi are *eukaryotic*. Prokaryotic cells are small and simpler in organisation than eukaryotic cells and do not possess membrane-bound organelles. Nevertheless bacteria may be classified in an orderly manner based upon their common properties. This classification (see Table 1.1) is best achieved by a method in which different bacterial isolates are compared for a number of characteristics and clustered according to percentage similarities. The main properties used in classification are as follows.

Morphology. One of the properties of major importance in classifying bacteria is the morphology of the individual bacterial cell, that is, whether it is rod-shaped (bacilli), roughly spherical (cocci) or spiral (spirochaetes).

Staining. Another property of major importance in classification is the reaction after staining by the method of Gram. Gram-positive organisms stain blue and Gram-negative organisms stain red. The difference is related to the relatively thick layer of peptidoglycan in the wall of Gram-positive organisms. Gram-negative organisms also have an extra bilipid membrane containing several important molecules such as lipopolysaccharide.

Physiology. The most important property is sensitivity to oxygen (obligate anaerobes), tolerance of it (aerobes) or requirement for it (strict aerobes).

Metabolism. The ability of an organism to utilise or produce a variety of biochemical substances; e.g. the fermentation of lactose by certain intestinal bacteria.

Chemical. Analytical techniques used to determine, for example, the type of fatty acid or respiratory quinone possessed by the organism.

Genetic. Genome size, guanosine and cystosine (GC) content, and DNA relatedness using hybridisation techniques.

The various clusters of organisms are given names according to a set of rules (*nomenclature*) and these names can be seen as a short code standing for a large set of properties of the organism. This allows people to communicate without reiterating the list of properties on each occasion. As new properties of bacteria are examined and new bacteria isolated and characterised, the clustering of the various organisms may change and so may the name, often to the confusion of the clinician.

Bacteria are named by a unique binomial comprising a generic name, e.g. Escherichia, always written with an upper case first letter, and a specific name, e.g. coli, always written with a lower case first letter. In publications the binomial name of an organism is italicised, e.g. *Escherichia coli*. The generic name may be shortened; e.g.

Table 1.1 Major groups of medically important bacteria.

Obligate intracellular bacteria: *Rickettsia, Chlamydia, Coxiella*

Free-living bacteria
  Mycelial growth: *Actinomyces, Nocardia, Streptomyces*
  Unicellular growth
    Absence of cell wall: *Mycoplasma*
    Presence of cell wall
      Flexible cell wall; motility by endoflagella (spirochaetes): *Borrelia, Treponema,*
          Leptospira
      Rigid cell wall; motility by exoflagella if motile
        Stain *red* by Ziehl–Neelson stain: Mycobacteria (all rod shaped)
        Stain *blue* by Gram stain (gram-positive)
          Spherical shaped, e.g. *Staphylococcus, Streptococcus, Enterococcus*
          Rod shaped
            Spore forming
              Anaerobic: *Clostridium*
              Aerobic: *Bacillus*
            Non-spore forming: *Corynebacteria, Listeria*
        Stain *red* by Gram stain (Gram-negative)
          Spherical shaped, e.g. *Neisseria, Moraxella*
          Rod shaped
            Spiral rods: *Spirillum*
            Curved or helical rods: *Vibrio, Campylobacter, Helicobacter*
            Straight, small rods: *Legionella, Brucella, Haemophilus, Pasteurella, Bordetella*
            Straight, large rods
              Fermentative metabolism, facultative anaerobes: *Escherichia, Salmonella,*
                Shigella, Yersinia, Proteus, Klebsiella
              Non-fermentative or oxidative metabolism, aerobic: *Pseudomonas*
              Fermentative metabolism, obligate anaerobe: *Bacteroides, Fusobacterium*

Staph. instead of *Staphylococcus*. Larger clusters of organisms, sharing fewer properties, are given Family names ending -aceae, e.g. Enterobacteriaceae, or Order names ending in -ales, e.g. Rickettsiales. These names do not imply a genetic relatedness but a phenotypic similarity.

For the purpose of epidemiological investigation it is useful to be able to *type* bacteria. This is the process of subdividing a species on the basis of one or a smaller number of properties. Some of the more important typing methods are:

Bacteriophage typing: determining the sensitivity of an isolate to a standard set of bacteriophages which are viruses that parasitise bacteria.

Biotyping: one or a small number of biochemical reactions that differentiate separate isolates of the same species into biovars (i.e. biological variants).

Serotyping: using specific antibodies raised against the cell wall or capsule to divide a species into serovars (i.e. variants defined by serological reactions).

Antibiograms: the use of patterns of antibiotic sensitivity and resistance.

Protein analysis: the use of the different patterns of whole cell, or cell membrane proteins, produced after electrophoresis or the different mobilities of enzymes after isoelectric focusing.

Plasmid typing: determination of the content and molecular weight of plasmids in a species.

DNA analysis: the use of restriction endonucleases and electrophoresis to produce different patterns based upon DNA fragment size, e.g. restriction fragment length polymorphism.

2 Ecology and Spread of Bacteria

The vast majority of micro-organisms have nothing to do with human disease. They live in soil, in water, in vegetable matter and in animals and plants. They fix nitrogen, degrade dead biological material, ferment fruit, producing alcohol and vinegar, give taste to foodstuffs, and so on. Only rarely do the organisms capable of causing infection in man actually do so.

NORMAL FLORA

Healthy animals are colonised with a normal flora of bacteria which is actually outside the intact animal since skin and mucous membranes form a barrier to the penetration of micro-organisms. The skin is a waterproof passive barrier but the gut mucosa is a highly active defence. When a patient dies gut organisms rapidly enter the circulation and are always found in post-mortem blood cultures. The only site in the body where there is a true epithelial-cell defect is in the gingival crevices adjacent to the teeth. This site is heavily colonised with a very large number of bacteria including fusiforms, peptostreptococci, spirochaetes and some aerobic streptococci. There is a continuous flow of polymorphonuclear leucocytes into gingival crevices which prevents invasion by the normal flora.

Normal commensals are those organisms always found in carrier sites in healthy individuals. The range of commensal flora is probably determined largely by the HLA type of the individual but Table 2.1 shows a typical distribution.

Occasional commensals are those organisms which are common in the environment and occasionally colonise patients without causing disease, e.g. *Pseudomonas aeruginosa*.

Pathogens are less commonly (e.g. *Staph. aureus*) or never (e.g. *Vibrio cholerae*) found as part of the normal flora, yet they have the ability to colonise, invade and cause disease.

Man's normal microbial flora protects him from invasion by other microbes (*colonisation resistance*). Minor changes in the environment may disturb the normal flora. Broad spectrum antibiotics clear away commensals and permit overgrowth of resistant organisms which are usually present only in small numbers (e.g. staphylococcal enterocolitis in patients given tetracyclines following gastric surgery and candida vaginitis following the use of ampicillin). Alternatively, new organisms

may invade a given site; e.g. colonisation of the mouth with coliforms within a day of giving a broad spectrum antibiotic.

THE SOURCE OF INFECTING BACTERIA

Normal and occasional commensals are not virulent and overt infection only results if the host defences of the patient are compromised and the organisms allowed to invade or contaminate an unusual site. These may be called *endogenous infections* and common examples are urinary tract infection due to *Esch. coli* (gut commensal), pneumonia due

Table 2.1 Typical normal flora.

SKIN

Normal skin	*Staphylococcus epidermidis*, alpha-haemolytic streptococci, *Acinetobacter* spp.
In skin folds with eccrine glands (e.g. axillae, groins)	*Staph. aureus*, coliforms in addition to above
Feet	As above plus coryneforms, fungi (yeasts and moulds)
People who do not wash	Mixed coliforms and anaerobes (faecal flora)

GUT

Mouth and oesophagus	Mixed anaerobes 10^9–10^{10}/g predominate over aerobic alpha- and non-haemolytic streptococci, etc. (ratio 10–100:1)
Stomach	Very few bacteria (low pH)
Upper small bowel	Predominantly mixed aerobes (coliforms, enterococci)
Lower small bowel	Anaerobes begin to predominate
Large bowel	Mixed anaerobes 10^{10}–10^{11} (*Bacteroides* spp., *Clostridium* spp.) predominate over aerobes (10–100:1)

RESPIRATORY TRACT

Nose:

Anterior vestibule	*Staph. aureus, Staph. epidermis*
Posterior part and nasopharynx	Alpha-haemolytic streptococci *Haemophilus influenzae* *Strep. pneumoniae*
Trachea and bronchi	Inhaled organisms being removed by mucociliary escalator
Alveoli	Usually sterile but for inhaled organisms undergoing active clearance

GENITOURINARY TRACT

Vagina	> 55 species mainly anaerobes, *Lactobacillus* spp. predominate
Endocervix and uterus	Organisms unusual
Urethra	Skin flora

---

to *Strep. pneumoniae* (upper respiratory tract commensal) and wound infections and abscesses due to *Bacteroides* spp. (gut commensal) following abdominal or pelvic surgery.

Although some infections arise from the overgrowth of, or invasion by, normal flora the majority arise as a consequence of the transmission of new species or strains, i.e. they are *exogenous infections*. Other infected persons either directly or indirectly are the most common source of infection for susceptible individuals. However, animals, especially domestic animals in the case of food poisoning, are another important source of infection. Occasionally the environment is the source of the invading organisms, e.g. *Pseudomonas aeruginosa*.

TRANSMISSION OF INFECTION

Bacteria are transmitted by the airborne route, by the faecal–oral route, via fomites (contaminated, inanimate objects) or by direct contact with infected patients or animals.

Airborne Route

Coughing and sneezing produce large mucous particles which tend to fall under gravity and fine particles aerosols (1–5 μm diameter) which remain suspended indefinitely in the air. These aerosols are important in the transmission of tuberculosis. Man-made aerosols containing the organism of Legionnaires' disease may be created from cooling towers and air-conditioning plants. On inhalation the aerosolised particles are not trapped in the upper respiratory tract but are deposited in the alveoli. This is the portal of entry for the infectious micro-organisms in the aerosols.

Viruses which infect the upper respiratory tract are best transmitted by direct inoculation on hands contaminated with the nasal mucus from an infected person. This may also be a method of transmission for *Strep. pneumoniae* and *Haemophilus influenzae*.

Faecal–Oral Route

A human carrier of a pathogen in the gut (e.g. *Salmonella typhi*) may inoculate food with the organism during preparation. Particularly dangerous foods are those which require a lot of handling and are not subsequently cooked, e.g. salads. Water supplies may also be contaminated by faecal material from infected humans.

The indirect faecal–oral route is important for highly infectious agents such as *Shigella sonnei* which readily contaminate the environment around a patient with profuse diarrhoea. It is easy to contaminate the hands and put the fingers in the mouth or eat before washing. Children transmit via this route very efficiently in nurseries and at school.

Foodstuffs may also be contaminated with organisms from the intestines of animals (especially poultry) during the preparation for sale in shops. Other foods may be contaminated with common environmental organisms, e.g. *Listeria*.

Direct Contact

Certain infections such as the sexually transmitted diseases require close direct contact with infected individuals. Other bacteria may enter a wound at the time of injury, e.g. *Clostridium tetani* as a consequence of a dirty injury. Transfer of organisms can also take place by the bite of an animal or an insect vector (e.g. the plague bacillus transmitted by rat fleas).

Fomites

Infections transferred via inanimate objects are very varied and include skin diseases from baths, intestinal infections from crockery and cutlery, septicaemia from intravenous fluids and wound infection from contaminated surgical instruments or dressings.

MICROBES AND DISEASES

There are relatively few microbes so specialised that they can live only at our expense. An example is *Neisseria gonorrhoeae* (the causative agent of gonorrhoea), pathogenic only to man. When such organisms are recognised in specimens from patients in the laboratory there is no problem in attributing a pathogenic role to them.

There are others which, while pathogenic only to man, are also capable of separate survival, e.g. *Salmonella typhi*, the causative agent of typhoid fever, which, while causing a distinctive human disease, may also be isolated from sewage, foodstuffs and water supplies. It is possible to isolate this organism in the laboratory from specimens both from patients with typhoid fever and from those who had the disease long ago. These latter, asymptomatic carriers of *S. typhi* are not ill, but represent a possible source of infection to others. It is logically quite possible for such a carrier to be suffering from another illness. This creates an obvious difficulty in the interpretation of laboratory findings.

Medical microbiology in industrialised societies has moved away from the specific, named fevers with a single pathogen (such as diphtheria, caused by *Corynebacterium diphtheriae* or plague caused by *Yersinia (Pasteurella) pestis*, or with only two or three pathogens (such as enteric fever cause by *S. typhi* or *S. paratyphi*, cholera caused by *Vibrio cholerae* or its El Tor biotype). These societies are now more concerned with respiratory infections, meningitis, and urinary tract infections which may be caused by many different types of organism, giving rise to very similar clinical features. In less prosperous societies, although these latter conditions also occur, the named fevers such as diphtheria, plague and cholera still play a large part in disease.

In the nineteenth century, the principles under which it was possible to assign a pathogenic role to an organism were clearly stated, and are known to us in modified form as Koch's postulates.

1. It must be possible to grow the organism in the laboratory.
2. The organism must be isolated from all cases of the disease.
3. It must not be isolated from those who do not have the disease.
4. It must be capable of causing the disease in a susceptible animal.

The principles were invaluable at the time of their introduction in finding out which organisms caused which disease. However, there were problems.

1. The leprosy bacillus (*Mycobacterium leprae*) and the organism of syphilis (*Treponema pallidum*) cannot be grown in the laboratory.
2. The pathogen cannot be isolated from all cases of human brucellosis (the diagnosis often depending on the identification of an antibody response to the pathogen).
3. As has already been indicated, asymptomatic carriage of *Salmonella typhi* (which under other circumstances causes typhoid fever) may occur.
4. Some human pathogens (e.g. leprosy bacillus) have susceptible animal hosts which are rarely available for laboratory work.

Even with the specific infectious diseases, therefore, the requirements of Koch's postulates often cannot be fulfilled completely and they can be complied with much more rarely in the non-specific infections which cause the majority of infectious conditions in developed countries.

Thus organisms cannot be arbitrarily divided into those which are, and those which are not, pathogenic to man. There is a spectrum from those organisms which are very virulent to those which cause disease only in occasional patients with compromised defences. There is almost no organism which cannot, on occasion, cause human disease. Moreover, the full range of infectious diseases in man has by no means been explored. Quite apart from the periodic recognition of 'new' tropical fevers (some with very disagreeable features such as Lassa fever, Marburg disease or Ebola fever), new infectious diseases of temperate climates continue to be revealed. Recent examples include the realisation that the commonest cause of infectious diarrhoea is *Campylobacter* enteritis, the emergence of *Clostridium difficile* as a cause of antibiotic-associated colitis, and the sensational appearance of a new form of pneumonia, Legionnaires' disease, caused by a newly recognised organism, *Legionella pneumophila*. The nature of the infectious diseases has changed, is changing, and will continue to change.

3 Pathogenesis of Bacterial Infections

The production of clinical disease due to bacteria is a consequence of a complex interplay of bacterial properties and host factors. *Colonisation* of the host is a prerequisite of any infection and this involves either attachment of bacteria to an epithelial surface or direct inoculation through a wound or by an arthropod. Virulent organisms have adhesions (e.g. the M protein fibrillae of *Strep. pyogenes*, adherence pili of *Esch. coli*) which promote attachment to cell-surface binding sites. The size of the inoculum is also important requiring that approximately 10^{10} organisms be ingested in the case of *Vibrio cholerae*, 10^6 in the case of *Salmonella* spp. and 10^{1-2} with *Shigella* spp.

After colonisation *local infection* may ensue at the portal of entry of the organism. Alternatively, organisms may *invade*, i.e. spread widely through the lymphatics or bloodstream and cause deep-seated infection. Common sites of invasion are the respiratory and gastrointestinal tracts. Some organisms produce disease at a distance without invading the host because they elaborate *toxins* (see below) which may also act locally or at a distance.

THE OUTCOME OF COLONISATION BY BACTERIA

Local Inflammation

After invasion pyogenic bacteria (e.g. *Staph. aureus, Strep. pyogenes*) characteristically induce local inflammation which explains many aspects (pain, erythema etc.) of the resulting disease. It is a complex process involving local accumulation of neutrophils, release of vasoactive peptides, prostaglandins (PGE_1 and PGF_2a), leukotrienes and 5-hydroxytryptamine, and consequent vasodilatation and oedema. The eventual result of this process is healing or abscess formation, necrosis and discharge of pus. Bloodstream dissemination of these pyogenic organisms (bacteraemia) may lead to multiple abscesses in any organ. (In a granulocytopenic patient, abscesses cannot form and the patient has mild inflammation without the formation of pus.)

Septic Shock

This syndrome is seen most often in hospitalised patients postoperatively or in intensive care. It involves diffuse arteriolar vasodilation with capillary leakage of

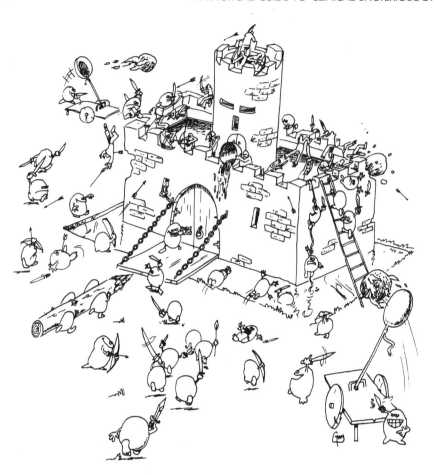

fluid and arteriolar obstruction by intravascular microcoagulation. Hypotension and tissue hypoxia result and particularly affect the kidneys and brain but often all important organs. A build-up of lactate from damaged tissue cannot be cleared by the damaged liver, resulting in acidosis. Oedema in the alveolar–capillary region leads to 'shock lung' or 'adult respiratory distress'. The syndrome is caused by a self-perpetuating cascade of inflammatory mediators, the primary ones being tumour necrosis factor and interleukin 6 produced by macrophages. The most efficient stimulus for septic shock is lipopolysaccharide (often called endotoxin) of the outer membrane of Gram-negative organisms (coliforms, *Pseudomonas* spp., *Neisseria* spp. and *Haemophilus* spp.), but the syndrome is sometimes seen with severe Gram-positive infections (*Staph. aureus*, *Strep. pyogenes*) and even after burns and severe trauma when no bacteria are detectable.

Granuloma Formation

Characteristically organisms which survive intracellularly grow slowly and stimulate a quite different response from the classic acute inflammatory abscess mediated by polymorphs. The pathological lesion in response to mycobacteria, *Listeria* and *Brucella*, for example, is the granuloma which consists of macrophages and lymphocytes. Central necrosis which is caseous or cheese-like can occur in tuberculosis. Granulomas may be found in any organ are also stimulated by foreign bodies and by agents which are as yet unidentified (e.g. Crohn's disease).

Toxins

Most bacteria make and excrete proteins (toxins) which are toxic to the host. In a few instances, a particular toxin is responsible for a characteristic illness. Some act locally whereas others are spread via the bloodstream and affect distant organs.

Enterotoxins

The best example is the toxin of *Vibrio cholerae*. The organism colonises the small intestine but does not invade. The toxin it produces has a two-part structure, one part 'B' resulting in binding to a specific ganglioside receptor of the mucosal cells, the other part 'A' splitting off and entering the cell to activate the adenyl cyclase system which controls the flux of Na^+ and Cl^- ions across membranes. As a consequence there are massive intestinal losses of water and electrolytes in cholera.

Clostridium perfringens and *Bacillus cereus* ingested in food may cause enteritis after sporulating and producing enterotoxins. Preformed toxins of different strains of *B. cereus* cause sudden acute vomiting similar to that seen after *Staph. aureus* food poisoning which is also due to a preformed toxin. *Cl. perfringens* type C toxin causes pigbel, an acute necrotising enteritis of the jejunum following a meal of pig meat first described in Papua New Guinea.

Tissue Necrosis due to Toxins

Cl. perfringens type A toxin and closely related toxins are responsible for the cell death associated with the spreading necrosis and oedema of gas gangrene. Alpha-toxin inactivates neutrophils at very low doses and higher concentrations act in concert with theta-toxin, collagenase and hyaluronidase under the anaerobic conditions present in dead tissue to cause spreading gangrene.

Staph. aureus and *Strep. pyogenes* possess a multitude of toxins which enhance their virulence. Important ones include *Staph. aureus* alpha-toxin which interferes with the permeability of cell membranes and enterotoxin F (or TSST-1) responsible for the toxic shock syndrome characterised by rash, hypotension, diarrhoea and desquamation. Streptolysin O binds to cholesterol in cell membranes and is cardiotoxic, streptolysin S is cytotoxic. Scarlet fever is a systemic febrile illness mediated by an erythrogenic toxin of certain strains of *Strep. pyogenes*. This organism also produces hyaluronidase and streptokinase which gives it the facility for moving through the superficial layers of the skin to cause erysipelas or through the fascial planes to cause necrotising fasciitis.

Neurotoxins

Tetanus results from the local production of a polypeptide by *Cl. tetani* growing under anaerobic conditions in a traumatic wound. The toxin enters motor nerves, travels up to the nerve endings and interferes with neurotransmission. Breakdown of the inhibitory action of the upper motor neurones on the motor reflex results in muscular spasms. By contrast, a neurotoxin which is preformed in canned and other foods contaminated by *Cl. botulinum* circulates after ingestion and causes flaccid paralysis.

Corynebacterium diphtheriae colonises the throat usually without symptoms. Strains which possess a phage coding for a toxin cause severe local disease (membrane formation and oedema with respiratory obstruction) and remote toxic effects particularly on heart muscle and nerves.

HOST DEFENCES

Protection against bacterial infection is achieved by the immune system. This is partly innate and partly adaptive, in that it is exposure to antigens which leads to the generation of specific antibodies and specific cell-mediated immunity. Increased susceptibility to infection occurs whenever there is a deficiency of one of the host defences, as shown in Table 3.1.

Many of the consequences of deficient host defences are best illustrated by the severe *inherited defects* included in Table 3.1. However, the list is almost endless and Table 3.2 gives further examples of some of the notable *acquired immunodeficiencies*.

Subversion of Host Defences

It can be seen from Table 3.2 that there are two critically important principles in combating the bacteria that constantly impinge on individuals. The first is the integrity of the mechanical barriers to infection and the second is the interaction between phagocytic cells, antibody and complement. Bacteria can do little about the first defence other than be opportunists but many bacteria possess virulence factors which interfere with the second mechanism (Table 3.3).

Table 3.1 Infections caused by deficiencies in host defences.

Host defence	Deficiency	Infection
Innate		
Physical barrier of skin	Burns	*Staph. aureus; Ps. aeruginosa*
Mucosal secretions	Cystic fibrosis	*Staph. aureus; Ps. aeruginosa*
Ciliary action	Chronic bronchitis	*Strep. pneumoniae; H. influenzae*
Commensal flora	Antibiotic therapy	Fungal, Gram-negative bacterial
Non-cellular killing	Complement factor deficiency	Fungal
Phagocytosis	Neutropenia; inherited disorders of functions	Severe Gram-negative sepsis; *Staph. aureus*
Adaptive		
Antibody response	X-linked hypogamma-globulinaemia	Pyogenic bacterial infection
Cell-mediated	Inherited disorders of number and function	Viral unless combined with antibody deficiency

Table 3.2 Effects of acquired immunodeficiencies.

Condition		Effect
Extremes of age	Infancy	Immature immune system
	Old age	Decreasing T-cell numbers and function
Diabetes mellitus	Ketoacidosis	Impaired neutrophil function
	Microangiopathy	Poor inflammatory response
	Neuropathy	Inadvertent trauma with ulceration
Operations	Clean	Break in integument
	Contaminated	Introduction of large numbers of bacteria
Splenectomy		Removes phagocytic filter in circulation
		Decreased B-cell precursors
		Decreased opsonisation
Renal disease	Acidosis	Impaired cell function
	Fluid overload	Risk of respiratory infection
	Dialysis	Break in the skin
	Steroids	Suppressed T-cell function
Chronic hepatitis		Disruption of fixed macrophages in liver
Malignancy	Lymphoma	Suppressed T-cell function
Organ transplantation	Drugs and irradiation	Neutropenia and suppressed T-cell function
	Hickman line	Portal of entry
Malnutrition	Protein, calorie or vitamin deficiency	Multiple defects

Table 3.3   Bacterial interference with host defence.

Avoidance of host defence	Organism
Failure to activate alternative pathway of complement	Certain *Esch. coli* and *Strep. pyogenes*
Inhibition of opsonising effect of antibody	Polysaccharides of many organisms (streptococci; *Neisseria*; *Haemophilus*)
Destruction of complement components	*Ps. aeruginosa*
Resistance of serum killing	*Neisseria*
Inhibition of chemotaxis	*Staph. aureus, Cl. perfringens*
Killing of phagocytes	Staphylococci; streptococci; clostridia; pseudomonads
Resistance to intracellular killing by macrophages	Mycobacteria; *Listeria*; *Brucella*

4 Laboratory Diagnosis of Bacterial Infections

When a patient is suspected of having a bacterial infection, microbiological investigation should be performed only if it is likely to help diagnosis and treatment or for epidemiological reasons. Appropriate specimens, properly collected, should be transported to the laboratory without delay. Sometimes a *rapid diagnosis* can be achieved by *microscopy* of the specimen and *antigen detection* methods but a *definitive result* must await *culture*, *identification* and *determination of the antibiotic susceptibilities* of the likely pathogen.

SPECIMEN COLLECTION

In some conditions, for example uncomplicated urinary tract infection, empirical treatment is usually successful and culture of urine from an outpatient could be reserved for those who fail to respond to standard therapy. Culture of urine from asymptomatic inpatients is rarely justified unless prior to cystoscopy. In other instances, the results of culture are often unhelpful, for example sputum culture in acute exacerbations of chronic bronchitis or in pharyngitis. In serious infections, it is always important to obtain specimens before starting treatment. The one exception is bacterial meningitis where immediate administration of benzylpenicillin before admission to hospital can be of great benefit.

The correct interpretation of results requires that all specimens must bear the patient's name and time of sampling and the request form must give details of the patient's age, ward or destination for the report, the symptoms and their duration, the site of the specimen, the investigation required and any antibiotic treatment. In addition the following guidance about various specimens should be followed.

- Aseptic collection is required for specimens which are normally sterile, e.g. cerebrospinal fluid (CSF) and blood. Blood cultures are often contaminated by skin flora unless scrupulous aseptic technique is observed and the needles are changed between patient and bottle. Only three sets need be collected, preferably as the temperature starts to rise, and they must be placed in a 37°C incubator soon after collection.
- Urine, even if it is a properly collected midstream sample, is usually contaminated with a few bacteria and these can multiply, producing false results, if transport is

CAN YOU SEE IT?

(MICROSCOPY)

(CULTURE)

CAN YOU GROW IT?

(SEROLOGY)

HAS IT LEFT A TRACE?

delayed and the specimen is not refrigerated. Bacteria may also die under these conditions leading to false-negative cultures.

- *Neisseria* spp. are delicate organisms so CSF with possible *N. meningitidis* and cervical and urethral swabs (rather than high vaginal swabs) for gonorrhoea should be transported urgently to the laboratory and cultured within hours.
- Sputum should be a deep-cough specimen, collected in the morning after

physiotherapy, and, except in the critically ill, is useless if the patient is receiving antibiotics.

- Pus is always superior to a swab in its yield of pathogenic bacteria and is essential for the isolation of mycobacteria.
- Wound swabs should be collected into transport medium and sent to the laboratory within 4 hours.
- Paired sera, with the first being collected as early in the illness as possible, are needed for serological investigation.

METHODS OF DIAGNOSIS

Microscopy, culture, bacterial identification and assessment of antibiotic sensitivities form the major part of the work of the clinical microbiology laboratory. In addition, detection of bacterial antigen can be attempted if a rapid result is needed, or if antibiotics have already been given. Finally, serological techniques may be used to detect past or recent episodes of infection when culture is impractical. Preliminary macroscopic examination of the specimen may be helpful since it will show if it contains pus or blood, if it is suitable for culture (a mucoid specimen of sputum is not) and very occasionally suggests the diagnosis, e.g. the presence of 'sulphur granules' in actinomycosis.

Microscopy

Examination of unstained films for the presence and nature of pus cells (CSF, urine) and the presence of bacteria is occasionally useful. Also, examination of stained films demonstrates the proportions and morphology of different bacteria (including those that subsequently fail to grow) and can give a rapid diagnosis in the case of purulent CSF or a suspected positive blood culture.

- Gram's stain provides the most important division of bacteria for both laboratory and clinical purposes.
- Ziehl–Neelsen stain is definitive for acid- and alcohol-fast mycobacteria but they also stain with auramine and fluoresce brightly under ultraviolet light, permitting scanning of a film at lower power when the organisms are scanty.
- Fluorescein-labelled antibodies to bacterial antigens may be used to diagnose infection (e.g. genital tract infection with *Chlamydia trachomatis*).
- Dark ground microscopy uses oblique illumination so that light is scattered from the bacteria and shows delicate organisms such as spirochaetes.

Antigen Detection

- *Immunoelectrophoresis* can be used to react pneumococcal antigen in sputum, urine or serum with specific antibody to give a line of precipitate in, for example, patients with pneumonia even if antibiotics have been given.

- *Enzyme immunoassay* detects bacterial antigen (e.g. *Chlamydia trachomatis*) when it binds to a solid phase coated with specific antibody and is incubated with antichlamydial antibody and then an enzyme-conjugated detector antibody. Addition of a substrate produces a colour reaction in proportion to the quantity of the antigen present in the specimen.

- *Polystyrene latex particles* coated with antibodies to bacterial antigens provide rapid and simple tests. The particles visibly agglutinate in the presence of the appropriate antigen. Antigens from *Streptococcus* group B, *Haemophilus influenzae* type b, *Streptococcus pneumoniae* and some strains of *Neisseria meningitidis* can be detected in CSF, serum or urine. Similar techniques are used in the identification of haemolytic streptococci and *Staph. aureus* grown *in vitro*.

- *Gas liquid chromatography* can be used to diagnose the presence of anaerobes in pus from abscesses in the abdomen, brain and elsewhere by detecting volatile fatty acids which are bacterial metabolic products.

- *DNA probes* have been used for the detection of *Legionella* spp. and enterotoxin production by *Esch. coli* but are not widely available.

Culture of Bacteria

Most pathogenic bacteria grow on artificial media, horse blood agar being the commonest, but some have more specific requirements. Specimens from normally sterile sites are usually inoculated onto blood agar and incubated in air enriched with 5–10% carbon dioxide and anaerobically with 10% carbon dioxide at 37°C. Most pathogens will grow in 48 hours. In cultures from sites with a normal flora, the chance of isolation of a pathogen can be improved by *selective culture* (Table 4.1).

Solid media demonstrate colonial morphology and the degree and purity of growth, and single colonies can be subcultured for biochemical identification. In the culture of urine, the number of bacteria present determines whether bacteriuria is thought significant ($>10^5$/ml). Liquid media, such as nutrient broth or Robertson's meat medium, help the isolation of bacteria present in small numbers or damaged by antibiotics, but give no indication of the numbers in the original specimen. Liquid media are also used in biochemical identification and as *enrichment cultures*, e.g. selenite F for *Salmonella* spp.

Identification of Bacteria

Colonies of most bacterial pathogens have a characteristic appearance and conditions for optimal growth. Selective media will aid identification, for example by showing lactose fermentation of Enterobacteriaceae, but Gram stain or other confirmatory tests are usually needed. Some tests can be performed the same day, for example the detection of the coagulase of *Staph. aureus*, the oxidase test for *Pseudomonas* species and

Table 4.1 Some solid media used for bacterial isolation.

Medium	Additive	Purpose
Blood agar (nutrient agar plus horse blood)	—	Primary isolation
Chocolate agar (heated blood agar)	—	Isolation of *Haemophilus influenzae*
Blood agar	Neomycin	Selection of anaerobes
Blood agar (Hoyle's)	Potassium tellurite, lysed blood	Selection of *Corynebacterium diphtheriae*
Cysteine-lactose electrolyte deficient (CLED) agar	Peptone, cysteine, lactose, bromothymol blue	Enterobacteriaceae: lactose fermenters turn yellow. Swarming by *Proteus* spp. inhibited
MacConkey agar	Peptone water agar, bile salt, lactose, neutral red	Enterobacteria: lactose fermenters turn pink. Swarming by *Proteus* spp. inhibited
Desoxycholate citrate agar	Sodium desoxycholate and citrate, lactose, neutral red	Selection of salmonellae and shigellae
Campylobacter agar	Growth supplement, lysed blood, vancomycin, polymyxin B, trimethoprim	Selection of *Campylobacter* spp. Incubate at 42°C in micro-aerophilic atmosphere
Löwenstein–Jensen	Glycerol, malachite green, salts, egg	Isolation of *Mycobacterium tuberculosis*

Campylobacter species, and motility and detection of catalase for *Listeria* species. Enteric pathogens can be identified provisionally by serological agglutination but confirmation by biochemical reactions is essential.

Other tests require incubation overnight. Pneumococci are distinguished from other streptococci by their sensitivity to optochin, enterococci blacken bile–aesculin agar, and *Haemophilus influenzae* grows on nutrient agar only near a filter disc containing X and V factors. A wide range of biochemical tests is easily performed by inoculating a commercial strip of cupules containing dried substrates and is most commonly used for the Enterobacteriaceae. The pattern of reactions produces a numerical profile identifying the organism (API system). When an outbreak occurs, the antibiotic sensitivity patterns and biochemical reactions of various isolates may point to a common source. Reference laboratories can further type organisms to distinguish similar strains, for example by showing lysis by bacteriophages (see Chapter 1).

Serology

Serological methods are used when the pathogen is difficult or impossible to grow but they usually require two specimens of serum separated by 10–14 days, so that diagnosis is retrospective. Infections due to *Brucella* spp., *Coxiella burnetti* and *Mycoplasma pneumoniae* are usually identified by a fourfold rise in titre between acute and convalescent sera tested at the same time. Sometimes a single high titre or detection of IgM antibody can give an immediate diagnosis. Diagnosis of syphilis requires three serological tests, each with different characteristics (see Chapter 16).

Significance of Results

Once a potential pathogen has been isolated or implicated, the clinical significance of the result must be assessed by the microbiologist. Information supplied on the request form is helpful but in serious infection there should be direct discussion between the microbiologist and the clinician. The microbiologist will often visit these patients. Coagulase-negative staphylococci isolated from blood cultures should not be dismissed as skin contaminants because they can derive from an infected central line or even infective endocarditis. Interpretation of isolates from sites with a normal flora requires knowledge of the standard of the specimen, the delay before inoculation onto media, the direct microscopy result, and whether growth is heavy or predominant. Gram-negative bacteria, for example *Pseudomonas* species, in the sputum of patients who have received antibiotics, are usually of little clinical significance except in ventilated patients. The antibiotic sensitivities of pathogens are withheld in cases where the clinical relevance of the specimen is in doubt and the clinician is requested to contact the microbiologist if necessary. Interpretation of serological results requires knowledge of the date of onset of illness, and the occurrence of any previous infection or immunisation.

5   Antimicrobial Agents

In the 1890s Paul Ehrlich predicted that chemical compounds could be found or synthesised that would specifically inhibit or kill a parasitic micro-organism without damage to the host. Such *synthetic compounds* have been found. By contrast penicillin is an *antibiotic*, a naturally occurring compound synthesised by one organism but capable in low concentration of inhibiting or destroying another (selective toxicity). The general term *antimicrobial agents* includes antibiotics and synthetic compounds such as the sulphonamides and trimethoprim. Bacteria are able to develop resistance to antimicrobials, and it is only by a combination of judicious use and ongoing intensive research that antimicrobial chemotherapy continues to be effective in most cases and to be one of the hallmarks of modern medical practice.

PRINCIPLES OF ANTIMICROBIAL CHEMOTHERAPY

- Decide if antimicrobial therapy is necessary. If the aetiology is unlikely to be infectious, or probably caused by agents other than bacteria, there is no point in giving an antibiotic. In general, bacterial infections should be treated early. Resolving infections do not usually require therapy.
- Choose an appropriate drug. The condition should be assessed both clinically and microbiologically. Take specimens for investigation before starting treatment. If treatment must commence immediately on clinical grounds, cover the likely bacteria with a narrow range antibiotic if possible.
- Give drug in adequate dose, at the correct intervals, and via an appropriate route. Underdosage is more common than overdosage, but neither is desirable. Potentially toxic drugs (e.g. the aminoglycosides) will require monitoring of peak and trough serum levels.
- Give drug for the correct duration. With few exceptions, antibiotics are prescribed for a course. This should extend for a few days beyond apparent clinical cure. Certain chronic infections, e.g. tuberculosis and osteomyelitis, will require prolonged treatment.
- Resist change without good reason. Antibiotics do not act instantly. Unless there is clinical deterioration, or the results of bacterial investigations dictate, the regimen should not be changed or added to.
- Only prescribe after asking about previous adverse reactions to antibiotics.
- Remove barriers to successful therapy, e.g. drain abscesses, remove foreign bodies.

CLASSIFICATION OF ANTIMICROBIAL AGENTS

Antimicrobial agents may be usefully classified by their mechanism of action, or by their spectrum of activity against different bacteria. (It is worth developing a working knowledge of both classifications as a useful guide to the selection of an appropriate agent.) The key to successful antibacterial action with little host toxicity is to target synthetic pathways which are unique to bacteria (i.e. the cell wall), or involve stages where the activity against the bacterial pathway can be selectively inhibited to a far greater extent than the similar host pathway (e.g. inhibitors of folic acid synthesis). Table 5.1 classifies commonly used antimicrobial agents by mode of action.

Table 5.1 Mode of action of commonly used antibiotics.

Target	Antibiotics
Cell wall	β-lactams (penicillins and cephalosporins), glycopeptides (vancomycin)
Cell membrane structure	Polyenes (antifungal agents, e.g. amphotericin B, nystatin), polymyxins (peptides)
Nucleic acids	(a) Indirect on folic acid synthesis (sulphonamides and trimethoprim)
	(b) Direct on DNA (quinolones) or on RNA (rifamycins, metronidazole)
Protein synthesis	(a) Inhibitors of ribosome 50S sub-unit (chloramphenicol, lincosamines, macrolides, fusidic acid)
	(b) Inhibitors of ribosome 30S sub-unit (tetracyclines, aminoglycosides)
Miscellaneous	Nitrofurans

SPECIFIC ANTIBACTERIAL AGENTS

Penicillins and cephalosporins differ from each other in that penicillins have a 5-membered thiazolidine ring, whilst cephalosporins have a 6-membered dihydro-thiazine ring fused to the β-lactam ring, but both are β-lactam antibiotics.

Penicillins

These antibiotics are bactericidal, inhibiting synthesis of peptide chains which cross-link the layers of glycans used in the formation of bacterial cell walls. They act best on dividing bacteria. Severe hypersensitivity may lead to anaphylactic shock; all penicillins are to be avoided in patients with a history of allergy to any one of them.

Benzylpenicillin (penicillin G) is active against Gram-positive bacteria, and Gram-negative cocci. Its activity is destroyed by β-lactamase. Its main clinical use is for streptococcal (including pneumococcal), staphylococcal (β-lactamase negative) and neisserial infections (meningococcal meningitis and gonorrhoea). It remains the drug of choice for clostridial gangrene and syphilis. Benzylpenicillin is inactivated by stomach acid and must be prescribed parenterally. Acid stable penicillins, e.g. phenoxymethyl penicillin (penicillin V), may be given orally.

Methicillin and *flucloxacillin* are stable against staphylococcal β-lactamase. Flucloxacillin is the treatment of choice for penicillinase-producing staphylococci. Resistance to methicillin coincides with resistance to flucloxacillin, and all cephalosporins. Oral and parenteral versions are available.

Ampicillin and *amoxycillin* are broad spectrum penicillins with activity against *Haemophilus* species, *Esch. coli*, some *Proteus* species and salmonella. These agents are not stable to β-lactamase. Their main use is in sensitive urinary and respiratory infections, enterococcal infections and listeriosis. Amoxycillin can be combined with clavulanic acid, an inhibitor of β-lactamase. The combination is active against many β-lactamase producing amoxycillin-resistant bacteria.

Ticarcillin, azlocillin and *piperacillin* are extended spectrum penicillins with activity against *Pseudomonas* species. Their main use is in the treatment of severe sepsis in which *Pseudomonas* species may be involved. They are often combined with a second drug, e.g. an aminoglycoside.

Cephalosporins

The site of action of these antibiotics is the same as that for the penicillins. They are less sensitive to β-lactamase activity. Note that 6–8% of patients hypersensitive to penicillin react to cephalosporins. All cephalosporins are inactive against enterococci. The *first generation* cephalosporins, e.g. cephradine and cephalexin, have been largely superseded by newer cephalosporins.

The *second generation* cephalosporins, e.g. cefuroxime/cephamandole/cefaclor, are characterised by good activity against *Haemophilus* species, and better stability against Gram-negative β-lactamase but they are less active against streptococci and staphylococci. Activity is good against β-lactamase producing *N. gonorrhoeae*. They are useful for post-operative sepsis (aerobes) and as second line therapy for chest infections. They are best given parenterally.

The *third generation* cephalosporins, e.g. cefotaxime and ceftazidime, have further improved β-lactamase stability, and some have activity against *Pseudomonas* species Cefotaxime is the drug of choice for neonatal meningitis, and is increasingly used for haemophilus meningitis in older children. Ceftazidime is widely used for empirical treatment of sepsis in immunosuppressed patients. It must be administered parenterally.

Tetracyclines

Tetracyclines (aureomycin/tetracycline hydrochloride/oxytetracycline/minocycline/doxycycline) are broad spectrum antibiotics which have been in use for over 40 years, although their usefulness is now limited by widespread and unpredictable resistance. They are bacteriostatic, acting on the 30S ribosomal sub-unit to prevent attachment of tRNA to mRNA receptor. They have good activity against *Chlamydia* species and mycoplasmas, and hence are widely used for non-gonococcal genital infection and atypical pneumonia. They are the antibiotic of choice for sensitive *Vibrio cholerae*, brucellosis, relapsing fever, rickettsial infections (e.g. typhus) and second line therapy for actinomycosis and syphilis. Tetracyclines are excreted via the gut and kidneys. They have a profound effect on normal gut flora, and diarrhoea is a frequent side effect as is candidal overgrowth of gut and female lower genital tract. Tetracyclines are taken up by developing bones and teeth and should be avoided in children, pregnancy and breast-feeding mothers (doxycycline should be used if a tetracycline is indicated).

Tetracyclines (except doxycycline) accumulate in renal failure and will exacerbate the condition. They are best prescribed orally, but not with milk or dairy products as tetracyclines will chelate with calcium and magnesium ions and fail to be absorbed (this is less significant with minocycline or doxycycline).

Macrolides (e.g. Erythromycin)

These act against the bacterial 50S ribosomal sub-unit, preventing elongation of peptide chain by inhibiting translocation. Their activity against Gram-positive aerobes is similar to benzylpenicillin and flucloxacillin and they are additionally active against chlamydiae, mycoplasmas (except *M. hominis*), *Legionella* species and *Campylobacter* species. They are useful as a substitute for penicillin and flucloxacillin in hypersensitive patients (e.g. upper respiratory and skin and soft tissue infections), also for atypical pneumonia, severe *Campylobacter* infections, non-gonococcal genital infection. They are the antibiotic of choice for diphtheria. Their toxicity is low, except for frequent gastrointestinal disturbance (vomiting) on oral administration (this problem may be reduced with new agents under development, e.g. clarithromycin and azithromycin) and they must be avoided in liver failure. There are oral and parenteral preparations.

Lincosamines (Clindamycin)

These have a similar action and spectrum of activity to macrolides. They are useful for inhalation pneumonia, and deep abscesses and are second line therapy for staphylococcal osteomyelitis; combined with an aminoglycoside they form a powerful combination for gut-associated sepsis. Their most notorious side effect is pseudomembranous colitis, caused by overgrowth of *Clostridium difficile*. There are oral and parenteral preparations.

Fusidic Acid

A steroid antibiotic with a mode of action similar to macrolides, this is a valuable drug in the treatment of severe staphylococcal infections. However, resistance develops owing to the selection of resistant mutants present in many strains of staphylococci and therefore fusidic acid should only be used in combination with another agent, e.g. a penicillin. Side effects include thrombophlebitis and reversible jaundice. It is available orally (the preferred route), parenterally and topically although the latter use should be discouraged in hospital.

Chloramphenicol

Chloramphenicol acts at the 50S sub-unit of the bacterial ribosome, blocking attachment of the growing peptide chain to a new amino acid. For most bacteria this action is bacteriostatic, although there is evidence to suggest that when treating meningeal infection with *Haemophilus* species, pneumococci or *Neisseria*, the action is bactericidal. Chloramphenicol has a broad spectrum of activity but its clinical use is restricted owing to possible toxicity. Idiosyncratic aplastic anaemia is the most important side effect, and is usually fatal. The 'grey baby' syndrome may occur in babies given high doses, resulting in total circulatory collapse. Blood and CSF levels should be monitored in children. Chloramphenicol should be avoided in hepatic disease and not be used for trivial infections. It remains a first line drug for typhoid fever (quinolones are an alternative), haemophilus meningitis (cefotaxime is an alternative), and brain abscess. It is available orally (the preferred route) or parenterally, as inactive esters which are hydrolysed to the active base. Chloramphenicol is used topically for infections of the outer eye.

Aminoglycosides

These are characterised by one or more aminosugar residues glycosidically linked to an aminocyclitol ring. Two groups are recognised, the streptomycins and the neomycin/kanamycin group. The precise mode of action is not understood, but involves inhibition of bacterial protein synthesis, and resistance may occur by a variety of mechanisms. Aminoglycosides are not absorbed by the oral route. Streptomycin is still used on occasion for mycobacterial infections but its use is now curtailed by widespread resistance and toxicity. Neomycin is available largely for topical use owing to renal toxicity and ototoxicity. It is occasionally given orally as a bowel decontaminating agent.

Gentamicin/Tobramycin/Netilmicin/Amikacin

In spite of their potential toxicity, these compounds continue to have an important role in the therapy of severe sepsis. Activity against Gram-negative aerobes including *Pseudomonas* species is very good. They show useful activity against staphylococci. Anaerobes and streptococci are resistant. They may however show synergy with β-lactam antibiotics in the treatment of difficult streptococcal infections, e.g. enterococcal endocarditis. They are frequently used in combination with an appropriate β-lactam for severe infections with Gram-negative aerobes and staphylococci. There is little to choose between gentamicin, tobramycin and netilmicin, although the latter is generally regarded as less toxic than the others. Amikacin is often active against bacteria resistant to the other agents, and should, therefore, be reserved for use in these circumstances. The antibiotics are given parenterally and serum levels should be monitored.

Glycopeptides (Vancomycin/Teicoplanin)

These are cell wall active antibiotics, inhibiting incorporation of the D-alanine dimer into the cell wall peptidoglycan. Permeability of the cell membrane and RNA synthesis are also adversely affected. They have a narrow range of activity against staphylococci and streptococci. They are generally bactericidal, except against the enterococcus, when the action is bacteriostatic. Vancomycin is used orally to treat *Cl. difficile*-associated pseudomembranous colitis. It is useful in severe staphylococcal or streptococcal sepsis, often in combination with gentamicin although it is potentially ototoxic and nephrotoxic, particularly when used with aminoglycosides. Vancomycin should be infused slowly over one hour and rashes and hot flushes of the head and neck region ('red man') may occur. Serum levels require monitoring.

Peptides (Polymixin/Colistin)

Peptides disrupt the osmotic integrity of bacterial cell membrane. Their activity is confined to Gram-negative aerobes, except members of the *Proteus/Providencia* group and *Serratia* species. They are neurotoxic and nephrotoxic and their use is now largely confined to topical preparations.

Folic Acid Synthesis Inhibitors

Sulphonamides are similar in structure to para-aminobenzoic acid (PABA), which is an essential component in the bacterial synthesis of folic acid. The synthetic pathway is blocked by the antibiotic, resulting in bacteristasis. In theory they are active against a wide spectrum of bacteria, including staphylococci, coliforms and to a lesser extent chlamydiae but widespread resistance limits their use now. Toxicity is usually confined to rashes and diarrhoea, rarely haemolytic anaemia or Stevens–Johnson syndrome (erythema multiforme). They should be avoided in the latter part of pregnancy, as they compete with bilirubin binding sites in serum, and may precipitate kernicterus in newborns. Oral and parenteral preparations are available.

Trimethoprim is an inhibitor of dihydrofolate reductase, preventing conversion of folic acid to folinic acid. It is also bacteriostatic with good activity against Gram-positive and Gram-negative aerobes, except *Pseudomonas* species. It is useful alone, or in combination with sulphamethoxazole for urinary tract, biliary and respiratory infections. Side effects are the same as for sulphonamides. Prolonged use may lead to megaloblastic marrow in the folate-deficient patient. Oral and parenteral preparations are available.

Co-trimoxazole is a fixed ratio combination of sulphamethoxazole and trimethoprim (5:1). Synergy of the two agents is often demonstrable *in vitro*, but its clinical significance has been challenged. It is useful for the same infections as trimethoprim and in bacterial dysentery and enteric fever. It has an important role in the treatment of *Pneumocystis carinii* infection. Side effects are almost always due to the sulphonamide component.

Quinolones

The mechanism of action of these antibiotics is complex, but primarily directed against synthesis of DNA by inhibition of DNA-gyrase activity. Side effects are generally mild (gastrointestinal and skin rashes) although CNS symptoms of varying severity have been recorded, ranging from dizziness to (rarely) fits. *Nalidixic acid* is essentially an oral urinary antiseptic, effective against enterobacteriaceae. Resistance develops easily. *Fluorinated 4-quinolones* are active against a wide spectrum of bacteria, particularly Gram-negative aerobes, including *Neisseria* species. Norfloxacin and enoxacin are orally active agents prescribed for urinary infections. Ciprofloxacin is highly active against *Salmonella* species including typhoid, and it is orally active against *Pseudomonas aeruginosa*. It is useful for the empirical treatment of severe sepsis, or when resistance to other agents is a problem. *Pseudomonas* species may develop resistance during therapy. It is highly active against *N. gonorrhoeae*, although resistance is being increasingly reported.

Other Agents

Rifampicin is a rifamycin, a complex macrocyclic molecule which inhibits RNA polymerase, and is bacteriostatic. Resistance readily develops owing to alteration of its target site and because of this, the drug should not be used alone during treatment. It has a very broad spectrum of activity, including staphylococci, streptococci, chlamydiae, *Legionella pneumophila*, neisseriae and *Haemophilus* species but is less active than other agents against enterobacteriaceae and *Pseudomonas* species and inactive against mycoplasmas. Its most important role is in the chemotherapy of mycobacterial infections as a component of combination chemotherapy. It may be combined with other antibacterial agents (e.g. trimethoprim), to treat difficult staphylococcal infections, e.g. *Staph. epidermidis* endocarditis or used as *single* agent for the *prophylaxis* of meningococcal and *Haemophilus* meningitis. Side effects include rashes and gastrointestinal disturbance, and a usually reversible hepatic dysfunction. Rifampicin reduces the efficacy of oral contraceptives. Urine and tears (and therefore contact lenses) will be discoloured orange during its use.

Nitrofurantoin is orally active against a wide range of urinary pathogens, including staphylococci, enterococci and enterobacteriaceae. Its precise mechanism of action is unknown. Resistance is becoming a problem; main side effects are gastrointestinal and skin rashes.

Metronidazole was originally introduced in 1959 for the treatment of *Trichomonas* infection; this compound was found by serendipity to be active against anaerobic bacteria. Its mechanism of action (which is bactericidal) is complex, involving reduction to an active metabolite within the bacterial cell. It is the drug of choice for most anaerobic infections. Side effects include metallic taste, interaction with alcohol causing vomiting, and a painful peripheral neuritis with prolonged courses. It is available orally and parenterally.

6 Sterilisation and Disinfection

Sterilisation is the complete removal or destruction of all living organisms on an item of equipment or in a solution. This includes bacterial spores. Sterilisation is required in clinical practice in situations in which instruments or materials are to be introduced into areas of the body that are normally sterile. Thus surgical instruments, intravenous infusions and drugs for injection must be sterile. There are four main ways of achieving this using *heat, radiation, filtration* or *chemicals*.

Disinfection is the reduction of the amount of micro-organisms in a given situation below that which is likely to cause infection. Vegetative organisms are killed but bacterial spores survive the process. In clinical practice certain utensils and surfaces cannot be regularly sterilised but potential pathogens need to be reduced to low numbers on floors and working surfaces, cutlery, bed pans etc. A special case is the cleaning of skin prior to injection or operation. *Heat* or *chemicals* can be used to disinfect.

STERILISATION

Heat

Moist heat kills organisms by coagulating proteins whereas *dry heat* kills by oxidation. The less moisture present the higher the temperature that is required to kill organisms. Thus there are two forms of heat sterilisation using either dry heat or moist heat in the form of steam.

Dry heat sterilisation uses a hot air oven operating at a temperature of 160–180°C with the objects being held in the oven for 20–60 minutes. Hot air oven sterilisation is used by pharmaceutical companies to sterilise ointments and in laboratories to sterilise glassware.

Steam sterilisation uses steam under pressure in an autoclave and operates at temperatures of 121 or 134°C with holding times of 15 minutes and 3 minutes respectively. Materials to be autoclaved are packed into special paper and placed in the autoclave chamber, which is sealed. (Solutions can be autoclaved by placing bottles of liquid with loosened caps in the autoclave.) All air is then removed using a vacuum pump so that pockets of air do not prevent the penetration of steam. High vacuum, high temperature autoclaves are in standard use in central sterile supplies departments

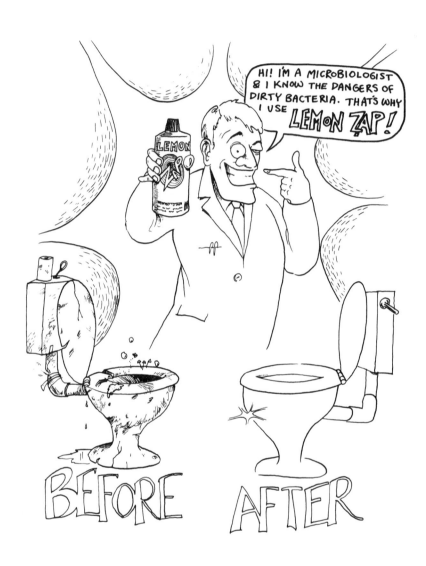

(CSSDs). Steam enters under pressure to achieve the high temperatures needed. The sterilisation process is monitored by measuring the temperature and pressure maintained during the whole cycle. Packs are sealed with unmarked tape which is impregnated with a thermochemical indicator. After autoclaving under correct conditions the tape is hatched by dark brown lines. Autoclaves are used to sterilise most surgical instruments, theatre gowns and surgical drapes. They are used in laboratories to sterilise media and also to make material microbiologically safe before disposal.

Simple downward displacement autoclaves, in which air is forced out by steam accumulating in the pressure vessel, are used to sterilise instruments in operating theatres. Instruments are laid on perforated trays and must not be wrapped. Quality assurance of this method of sterilisation is difficult.

Radiation

Sterilisation by radiation kills micro-organisms by producing highly reactive free radicals within bacteria thus disrupting many intracellular molecules including DNA. Gamma-irradiation is used to sterilise objects and the dose is usually 25 kiloGrays. Radiation is used commercially to sterilise relatively heat-sensitive materials such as surgical gloves, needles, syringes, infusion sets and some prostheses.

Filtration

Sterilisation by filtration using membrane filters with a pore size of 0.22 mm is the method of choice for liquids such as blood products and various pharmaceuticals which are heat labile. In strict terms, filtration does not sterilise as it does not remove viruses.

Chemical Sterilisation

Only two chemicals are commonly used to sterilise objects, ethylene oxide and formaldehyde. Both kill micro-organisms by alkylation of proteins and nucleic acid and both are applied by introducing them into a chamber containing the objects to be sterilised. The process is controlled by using a biological indicator (a known quantity of micro-organism introduced into the chamber with each load). With ethylene oxide a chemical indicator can be used to show that an item has been processed.

Ethylene oxide is used to sterilise heat-sensitive materials such as endoscopes and some prostheses. Formaldehyde is most commonly used to decontaminate bulky medical apparatus such as anaesthetic equipment and occasionally for endoscopes.

DISINFECTION

Heat

Heat disinfection is best known as it is applied to milk in the process of pasteurisation. Heat is also used in hospitals to disinfect such items as anaesthetic tubing, crockery, cloths and bedpans at a temperature of 82°C.

Chemicals

Chemical disinfectants are a collection of chemicals with different antibacterial spectra and uses. (Antiseptics are disinfectants which can be applied to skin or wounds.)

Phenolics are broad spectrum chemicals used for disinfecting inanimate objects such as floors.

Chlorhexidine has a narrow spectrum of activity (more against Gram-positive than Gram-negative bacteria) but is relatively non-toxic and is used for surgical disinfection.

Aldehydes such as *glutaraldehyde*, are broad spectrum agents used for disinfecting endoscopes but are toxic to users, so they must be handled with care.

Halogens are good broad spectrum disinfectants with chlorine in the form of *hypochlorite* being used for hard surface disinfection (but not on metal) and *iodine* or *iodophors* being used as a skin disinfectant.

Alcohol has a broad spectrum and in the form of isopropyl alcohol is used to disinfect clean surfaces (thermometers) and as a rapid skin disinfectant.

Quaternary ammonium compounds (e.g. cetrimide) are antimicrobial but have no effect on bacterial spores, mycobacteria and encourage the growth of pseudomonads. They are commonly used for skin cleansing (other than in operating theatres) since they also have detergent properties.

REMEMBER: Articles must be clean before they are sterilised or disinfected as dirt diminishes the efficiency of sterilisation and may inactivate disinfectants. Make up disinfectants freshly prior to use and discard excess. Nosocomial infection may be caused by contaminated disinfectants.

7 Immunisation Against Bacterial Infections

The control of both viral and bacterial communicable diseases remains an important goal for society. The purpose of immunisation is to prevent infection and the potential success of such programmes has been well demonstrated by the global eradication of smallpox. The maintenance of high levels of herd immunity is critical in preventing common infections. To this end it is particularly important that the facts concerning immunisation and its benefits and risks are presented clearly to the target population. No vaccine is entirely safe, but misinformation concerning a particular vaccine may lead to a poor uptake owing to fear of injury, particularly to a healthy child. Equally undesirable is complacency, which may also lead to a dangerous lowering of herd immunity. Continuous surveillance of the efficacy of vaccines and immunisation programmes is necessary to assess (a) side effects, (b) decreasing efficacy (owing to change in antigenicity or prevalence of a particular serotype), and (c) the level of uptake in the community. In addition, surveillance will assess the role of other methods of disease control (e.g. antibiotic use, public health measures).

PASSIVE IMMUNISATION

Passive immunisation involves the injection of human immunoglobulin. The protective effect is apparent within 24 hours but lasts for only a few months. Normal human immunoglobulin is obtained from pooled plasma from healthy blood donors. It contains adequate amounts of antibody against common viral infections, e.g. measles and hepatitis A, but it has no place in the prevention of bacterial infections. Specific human immunoglobulin is prepared from high-titre pooled plasma from convalescent donors, or individuals recently actively immunised against the relevant organism. Among bacterial infections specific human immunoglobulin has an important role in the management of individuals following exposure to tetanus and in diphtheria. All human blood products are prepared from donors free of infection with hepatitis B or C, and the human immunodeficiency virus (HIV).

ACTIVE IMMUNISATION

Active immunisation involves the administration of inactivated or attenuated organisms or their products (Table 7.1). In general, bacterial vaccines work by stimulating the production of circulating antibody, with BCG being the important exception. A minimum of two doses is required for primary immunisation except in the case of BCG (single dose). Following the first dose a slow, primarily IgM response occurs. Second and subsequent doses lead to a more rapid IgG response. High levels of antibody may persist for years, but even if the level falls, restimulation (either by exposure or a booster dose) will result in a rapid response.

Table 7.1 Examples of types of bacterial vaccine.

Live attenuated	Bacille Calmette–Guérin (BCG)
Inactivated	Pertussis, cholera, typhoid
Product	
Toxoid	Diphtheria, tetanus
Capsule	Meningococcus (A and C), pneumococcus
	Haemophilus (type b)
Culture filtrate	Anthrax

ADMINISTRATION OF BACTERIAL VACCINES

Most bacterial vaccines are given subcutaneously or intramuscularly, often with adjuvants, except BCG which is always given intradermally. Adjuvants are substances which enhance the antibody response. Aluminium phosphate and aluminium hydroxide are common adjuvants found in preparations of diphtheria/tetanus (DT) and diphtheria/tetanus/pertussis (DTP) vaccines. The use of combinations of bacterial vaccines may enhance the efficacy of the components. Thus pertussis vaccine acts as an adjuvant for diphtheria toxoid. Hib vaccine contains the polyribosyl-ribitol polysaccharide (PRP) of the capsule of *Haemophilus influenzae* type b conjugated to bacterial proteins to increase its immunogenicity.

There are contraindications to immunisation but care should be taken not to regard too many minor conditions as contraindications in case the level of uptake in the population is lower than necessary to control a disease effectively. The following are general contraindications.

- Severe acute illness.
- The use of live vaccines in pregnancy.
- Corticosteroid therapy, immunosuppressive treatment including radiation, malignancy and the use of live vaccines.
- Live vaccines within 3 months of injection of human immunoglobulin.
- BCG in persons infected with HIV.

SCHEDULES FOR MASS IMMUNISATION

Immunisation schedules for infants and children vary with the needs of particular populations. They are continually being reassessed and modified. A medical practitioner should be familiar with the local policy. Table 7.2 lists the schedule currently widely used in the UK (*Immunisation Against Infectious Diseases* (1992) HMSO, London).

Table 7.2 Schedule for immunisation of infants and children.

Vaccine	Age
Diphtheria/tetanus/pertussis, polio vaccine and Hib	3 doses at 2, 3 and 4 months
Measles/mumps/rubella	Minimum age 12 months (usually 12–18 months)
Booster diphtheria/tetanus/polio	4–5 years
Rubella	10–14 years (girls only if missed at 12–18 months)
BCG	10–14 years (infancy if indicated by local policy)
Booster tetanus and polio	15–18 years

Whooping Cough (Pertussis)

Whooping cough is particularly liable to result in severe complications and death in infants under 6 months of age. The incidence of the disease began to fall in the 1950s before the onset of widespread immunisation. The introduction of the vaccine accelerated this decline. Widespread public anxiety over the safety and efficacy of the vaccine in the early 1970s lead to a decline in uptake to below 30%. Major epidemics occurred in 1977/78 and 1981/83. Subsequently vaccine uptake has increased to over 70% and this is reflected in the current low levels of disease notification.

The vaccine is a suspension of killed *Bordetella pertussis* containing the three most important agglutinogens, and is usually given as a component of a triple vaccine combined with diphtheria and tetanus, with aluminium hydroxide as an adjuvant.

Swelling and redness may occur at the injection site, and screaming and fever may occur with triple vaccine. There are a small number of reports of severe neurological complications following pertussis vaccine, but the role of the vaccine in these complications is uncertain.

In addition to the general contraindications, pertussis vaccine should not be given to children who suffered severe local or general reactions to a previous dose.

Diphtheria

Diphtheria is now rare in the UK. Most cases are imported, but maintenance of immunity levels is essential to prevent outbreaks. Damage to the myocardium, nervous tissue and adrenals is toxin-mediated, hence the use of anti-toxin for passive immunity in acute cases.

Diphtheria vaccine is a formalinised preparation of purified toxin (toxoid). An adjuvant is required, usually aluminium phosphate or hydroxide, or combination with pertussis vaccine. Diphtheria vaccine is given as a component of the triple vaccine to infants (Table 7.2). Booster doses are required at school entry, and after contact with cases or carriers. For adults a low-dose, more purified preparation is used to avoid severe reactions in persons already immunised.

Tetanus

The spores of *Clostridium tetani* are widespread in nature, and any injury, particularly if contaminated with soil, represents a risk of infection. Consequently, it is in every individual's interest to establish immunity to tetanus.

Tetanus vaccine is prepared as a formalinised cell-free preparation of toxin (a toxoid). In common with diphtheria toxin, it is a poor immunogen and adjuvants are necessary to improve efficacy as described above. Tetanus toxoid is combined with diphtheria and pertussis for the primary vaccination of infants. For active immunisation after injury adsorbed tetanus vaccine should be used. Swelling and redness at the site of injection are not uncommon and may persist for some days. More generalised reactions are rare. Further doses should not be given to individuals who suffered a severe local reaction to a previous dose.

Tetanus prophylaxis does not detract from the need for thorough debridement of the wound. If the wound is more than 6 hours old, contains devitalised tissue or is contaminated with soil or manure, specific tetanus antiglobulin should be given, followed by a reinforcing dose of vaccine. If the patient has not been previously immunised, or the status is unknown, a full course should be given.

Tuberculosis (BCG)

The incidence of tuberculosis declined until the mid 1980s. Now infection in the UK is associated with elderly patients and the itinerant or immigrant populations.

The vaccine is an attenuated strain of *Mycobacterium bovis*, the bacille Calmette–Guérin (BCG). Vaccine is administered intradermally, and never intramuscularly or subcutaneously. Within 2 to 6 weeks a small indolent papule appears which may develop into a shallow ulcer. The lesions persist for some weeks before healing.

With the exception of newborn infants, vaccination is always preceded by tuberculin testing. Persons who react to tuberculin do not require vaccination.

SELECTIVE IMMUNISATION

Cholera does not occur in most parts of the world but travellers to some countries require vaccination against the causative organism. The vaccine is a phenol-preserved heat-killed suspension of *Vibrio cholerae*. Protection is effective for about 6 months.

Typhoid vaccine is also given to travellers. The monovalent vaccine is derived from phenolised heat killed *Salmonella typhi*. Two doses are required for full protection, over 4 to 6 weeks. Protection (70–80%) lasts 3 years or more. A single dose typhoid vaccine consisting of the Vi polysaccharide antigen of *S. typhi* and an oral typhoid vaccine are also now available. There is no effective vaccine against paratyphoid.

Anthrax is rare in industrialised countries in which occupational exposure to animal skins, wool and bone is the main indication for immunisation. Human anthrax vaccine is an alum precipitate of the sterile filtrate from the Sterne strain of *Bacillus anthracis*. Frequent boosters are necessary because antibodies decline rapidly.

Neisseria meningitidis can be divided into antigenic groups. The commonest are A, B, C, Y and W135. Human meningococcal vaccine is a purified extract from the capsular polysaccharide of groups A and C. There is no effective vaccine against group B strains. In some countries such as the UK the majority of infections are caused by group B strains. The vaccine is therefore of limited use in such countries. Travellers to areas where groups A and C are prevalent (e.g. the sub-Saharan meningitis belt in Africa) should be vaccinated.

Other Vaccines

A pneumococcal vaccine containing the capsular polysaccharide of the more common antigenic types is indicated for splenectomised patients or those with functional asplenia due to sickle cell anaemia. Protection is of the order of 80%, and lasts for up to 4 years.

8 Clinical Syndromes

The clinical syndromes caused by bacterial infection reflect (1) local replication in an organ system or (2) general invasion and dissemination, e.g. septicaemia, or (3) the remote effects of toxins produced by bacteria, e.g. tetanus, or (4) immunologically based disease, e.g. rheumatic fever. The most important organisms for each clinical syndrome are listed in this chapter and detailed descriptions of each organism are given in subsequent chapters. When presented with an infected patient, it is important to have a general idea of the organisms which may be causing an infection, so that appropriate chemotherapy can be chosen before the results of cultures are available.

UPPER RESPIRATORY TRACT INFECTIONS

Bacterial upper respiratory infections usually occur with organisms of relatively low virulence which are normal commensals in the nasopharynx. These grow well in obstructed cavities. Obstruction to sinuses and the middle ear usually arises because of the inflammation and oedema caused by upper respiratory virus infections which are exceedingly common.

Otitis media	*Streptococcus pneumoniae*; non-capsulate *Haemophilus*
sinusitis	*influenzae* alpha-haemolytic streptococci
Otitis externa	Usually allergy but may be a wide variety of organisms including *Pseudomonas aeruginosa* secondary to local steroids or antibiotics. *Staphylococcus aureus* may cause boils in external auditory canal
Acute epiglottitis	Capsulate *H. influenza* type b
Acute periorbital cellulitis	Capsulate *H. influenzae* type b (children); *Strep. milleri* (adults)
Chronic sinusitis	Doubtful bacterial aetiology
Chronic ethmoid sinusitis	Epithelial hypertrophy, debris and mixed anaerobes
Acute pharyngitis	*Strep. pyogenes* (viruses very common)
Tonsillitis	*Strep. pyogenes* (Epstein–Barr virus). Also candida in the immunosuppressed
Whooping cough	*Bordetella pertussis*

LOWER RESPIRATORY TRACT INFECTIONS

Classical lobar pneumonia is characterised by local chest pain, high fever, rigors, high peripheral white cell count and dry cough. The most common predisposing event is an acute virus infection (e.g. influenza) during the 3 weeks before the onset of pneumonia. The chest X-ray will probably be normal initially but will then show lobar or segmental shadowing.

Bilateral basal or 'broncho'-pneumonia is often associated with congestive cardiac failure and is a preterminal event in many elderly patients. *Strep. pneumoniae* and *H. influenzae* are commonly isolated from the sputum but the role of these organisms in pathogenesis is not well understood.

'Atypical pneumonia' is the name given to cases of pneumonia which are clinically and radiologically different from classical pneumonia, are caused by a different range of organisms and are diagnosed by serology rather than culture of the causative organisms.

Lobar pneumonia	*Strep. pneumoniae*
Bronchopneumonia	*Strep. pneumoniae*
	H. influenzae
Atypical pneumonia	*Mycoplasma pneumoniae*
	Legionella pneumophila
	Coxiella burnetti
	Chlamydia psittaci
	Chlamydia pneumoniae

Exacerbations of chronic obstructive airways disease are precipitated by virus infections (particularly rhinovirus) and manifest as increased cough, sputum, breathlessness and fever. The usual isolate from the sputum is *H. influenzae* and this is often treated unnecessarily with antibiotics.

Cystic fibrosis leads to colonisation of the lower respiratory tract with *Staph. aureus* or *Ps. aeruginosa*. Again epidemiological evidence suggests that exacerbations are precipitated by viral infections allowing these organisms to flourish.

Inhalation of stomach contents during acute collapse or during anaesthesia, or inhalation of a foreign body, will lead to basal pneumonia, often on the right side (the right main bronchus being the most likely bronchus to be involved) and will result in proliferation of oral flora, particularly mixed anaerobes. These lesions often cavitate causing large abscesses with fluid levels visible on chest X-ray.

In patients ventilated on the intensive care unit, colonisation of the upper respiratory tract and the trachea with unusual organisms such as *Ps. aeruginosa*, *Klebsiella* species, *Enterobacter*, species or even *Esch. coli* is very common. These are known as nosocomial (hospital-acquired) infections. Lower respiratory infection is difficult to diagnose but unexplained fever and tachycardia with new shadows on chest X-ray are an indication to treat the organisms isolated from tracheal aspirates.

GASTROINTESTINAL TRACT

Stomach

Gastritis and duodenitis may be caused by *Helicobacter pylori*.

Gall Bladder

Acute cholecystitis is usually associated with obstruction to the flow of bile and stones and ascension of normal duodenal flora up the biliary tree. The organisms commonly isolated are therefore coliforms and enterococci. Acute cholangitis sometimes with liver abscess formation is also associated with obstruction (often due to a carcinoma) and may follow endoscopy. Liver abscesses may also result from blood-borne spread.

Acute cholecystitis   Coliforms, enterococci
Acute cholangitis   Coliforms, enterococci
Liver abscesses   Coliforms, enterococci
                         Anaerobes often *Strep. milleri*

Appendix

Acute appendicitis arises because of obstruction of the neck of the appendix remnant. The normal colonic flora proliferates within the appendix. Certain organisms predominate including anaerobes such as *Fusobacterium* species.

Large Bowel

In acute enteritis the most important pathogenic mechanism is gut epithelial cell destruction with local invasion although there may be some toxin-mediated damage. Certain species of *Salmonella* (e.g. *S. typhimurium, S. hadar*) may invade further and cause bacteraemia but *Campylobacter* and *Shigella* very rarely do so. Outbreaks of enteropathogenic *Esch. coli* occur in institutions such as hospitals and schools generally affecting children but occasionally staff and teachers as well. Dysentery is diarrhoea with blood and mucus in the stools classically due to *Shigella flexneri* or *Shigella dysenteriae* or the parasite *Entamoeba histolytica*. However, the classical forms are rare in northern Europe but there may be blood in the stools in the common forms of acute enteritis (particularly with *Campylobacter* species).

Cholera, post-antibiotic colitis, clostridial and staphylococcal food poisoning are all toxin-mediated diseases. Neither these nor uncomplicated acute bacterial enteritis benefit from antibiotic treatment.

Acute enteritis   *Campylobacter jejuni*
                         Salmonella species
                         Shigella species
                         Some *Esch. coli* strains
Classical dysentery   *Entamoeba histolytica*
                         Shigella species
Cholera   *Vibrio cholerae*
Post-antibiotic colitis *Clostridium difficile*
'Food poisoning'   As acute enteritis plus *Staph. aureus, Cl. perfringens* and
                         Bacillus cereus

CENTRAL NERVOUS SYSTEM

Meningitis

Acute bacterial meningitis is characterised by headache, fever, photophobia, neck and back stiffness (nuchal rigidity) and mental confusion although these classical

Table 8.1   Meningitis—diagnosis.

	Normal	Viral	Pyogenic	TB
WBC/mm³	<5	20–200	>>200	>200
Predominant cell type	—	Lymphocytes	Neutrophils	Lymphocytes
Protein (g/l)	<0.45	1.0	>2.0	2.0
CSF/Blood glucose	59–80%	Normal	Low (<50%)	Very low (<<50%)
Organisms	—	None	Often	40–80%

Age	*Organism*
Neonates	*Esch. coli*
	Group B streptococcus
	Listeria monocytogenes
Infants	*H. influenzae* type b
	Strep. pneumoniae
	Neisseria meningitidis
Adults	*N. meningitidis*
	Strep. pneumoniae

presenting features may be absent in a neonate or elderly patient. Untreated, the patient will become comatose then die. Meningitis is diagnosed by examining the cerebrospinal fluid (CSF) for surrogate markers of infection and for the organisms themselves. Table 8.1 provides a guide to the indications given by CSF test results.

The likely organisms causing meningitis depends on the age of the patient (as shown in Table 8.1). Tuberculous meningitis is a 'post-primary' manifestation of bloodstream spread of *Mycobacterium tuberculosis*, common in areas of high endemicity. After head injury or neurosurgery, meningitis is often due to invasion by commensal flora from the upper respiratory tract. Although commonly due to *Strep. pneumoniae*, the organism involved is difficult to predict.

Abscess in the Nervous System

Brain abscesses present as space-occupying lesions spreading from local sepsis (e.g. chronic mastoiditis) or being seeded from distant sources (e.g. infective endocarditis) although the source is most often obscure. Extradural abscesses also arise by seeding from the arterial circulation or by extension from osteomyelitis.

Brain abscess	Mixed anaerobes, alpha-haemolytic streptococci
Extradural abscess	*Staph. aureus*

Infections of the heart

Bacterial infections of the endocardium are relatively common compared to those of the myocardium and pericardium. Where tuberculosis is common pericarditis may occur in post-primary or reactivation disease. In infective endocarditis a damaged valve causes turbulent flow which allows the deposition of platelets and the formation of platelet thrombus. This may become colonised with bacteria passing in the bloodstream. These bacteria are commonly streptococci of the oral flora. Intravenous drug abusers and individuals with prosthetic valves become infected with other organisms, particularly staphylococci.

Endocarditis
    Patients with damaged valves   Alpha-haemolytic streptococci
    Intravenous drug abusers   *Staph. aureus*
    Prosthetic valves   *Staph. epidermidis*
Pericarditis
    Constrictive due to fibrosis
    of healed lesion   *M. tuberculosis*
    Following heart surgery   *Staph. aureus*
        Staph. epidermidis

Myocarditis
    2° to overwhelming   *Staph. aureus*
    septicaemia   *Strep. pneumoniae*

RENAL TRACT

Urinary Tract Infection

Significant bacteriuria is defined as $> 10^5$ CFU (colony-forming units) of one organism per millilitre in two separate specimens of fresh and cleanly collected urine. The microbiological definition is fulfilled in about half of patients complaining of pain, urgency or frequency of micturition. It is also found in the urine of many asymptomatic patients but has important implications only in a small number of clinical situations (e.g. infants where structural abnormalities are common and pregnancy in which subsequent pyelitis is common). Isolates from the urine are as shown in Table 8.2.

Pyelonephritis results from organisms ascending the ureters from the bladder. Patients will have loin pain and signs of systemic infection. Repeated infection in infancy may cause loss of the renal parenchyma resulting in scarring. A small shrunken kidney with poor function is called 'chronic pyelonephritis' but this does not imply active infection.

Table 8.2 Bacteria in urine.

	Hospital (%)	General practice (%)
Esch. coli	48	68
Enterococcus faecalis	11	6
Staph. saprophyticus	10	5
Klebsiella, Enterobacter, Serratia	9	6
Proteus species	5	4
Pseudomonas species	5	—
Other	12	11

GENITOURINARY INFECTIONS

The most common syndromes of genitourinary infections are (a) urethral discharge usually with pain on micturition and (b) genital ulceration often with inflamed lymph nodes (buboes) in the groin. In women, symptomatic or asymptomatic urethral infection may ascend into the fallopian tubes and to the peritoneum to cause pelvic inflammatory disease characterised by lower abdominal pain with fever. Following genital ulceration the only generally invasive condition is syphilis in which the local lesion heals spontaneously after 4–6 weeks.

Acute urethritis	*Neisseria gonorrhoeae*
	Chlamydia trachomatis
Genital ulteration	
Chancre	*Treponema pallidum*
Chancroid	*Haemophilus ducreyi*
Lymphogranuloma venereum (LGV)	*Chlamydia trachomatis (serotypes L1, L2, L3)*
Granuloma inguinale	*Calymmatobacterium granulomatis*

SKIN AND SOFT TISSUE

Group A streptococcus with concomitant *Staph. aureus* infection causes impetigo—a superficial blistering, ulcerating and crusting lesion around the mouth. A slightly deeper infection in the skin is erysipelas with bright red inflammation and oedema which spreads bilaterally across the face. Deeper still, the organism may cause necrotising fasciitis. The infections, particularly impetigo which causes outbreaks in schools, are highly infectious to others by direct contact.

Superficial abscesses (boils, carbuncles)	*Staph. aureus*, sometimes mixed anaerobes and streptococci

Superficial ulcers	Often colonised with coliforms, anaerobes and Gram-positive cocci which do not need treatment
Superficial ulcers and spreading cellulitis	Beta-haemolytic streptococci
Impetigo	*Strep. pyogenes* and *Staph. aureus*
Erysipelas (bright red rash with oedema)	*Strep. pyogenes*
Necrotising fasciitis	*Strep. pyogenes* and *Staph. aureus*

BONES AND JOINTS

Acute infections of bones and joints may arise spontaneously following blood-borne spread, may follow compound fracture or penetrating injury or may arise post-operatively, particularly in association with prostheses.

Acute osteomyelitis	
commonly	*Staph. aureus*
in sickle cell anaemia	*Salmonella* species
following trauma	*Staph. aureus*, coliforms, anaerobes
post-operatively	*Staph. epidermidis*
as part of systemic disease	*M. tuberculosis; Brucella* species
Pyogenic arthritis	*Staph. aureus*
	Strep. pyogenes
	N. gonorrhoeae

SYSTEMIC INFECTIONS

Predisposing factors are given in Table 8.3. Some bacterial infections are characterised by a phase of bacteraemia. The features of bacteraemia are intermittent high fever, rigors with leucocytosis or leucopenia. Sepsis implies bacteria multiplying in the bloodstream and septic shock arises when the acute phase cascade is triggered by Gram-negative cell membrane lipopolysaccharide or by some undetermined factor in Gram-positive organisms.

Fever in Travellers

Bacteraemia is an important feature of the pathogenesis of a number of infections that are rare in the UK, e.g. typhoid fever, brucellosis, relapsing fever and leptospirosis. Blood cultures and microscopy must be performed on febrile patients returning from abroad (*NB: remember malaria.*)

Table 8.3 Systemic injections.

	Predisposing factors	Organisms
Hospitalised patients	Intravenous cannulae	Skin commensals
	Deep-seated wound infections	Coliforms, anaerobes
	Urinary tract infection	As urinary tract infection
	Neutropenia	Gut organisms
Community acquired	Renal tract infection ⎫ Gall bladder infection ⎭ Chest infection	*Esch. coli* or enterococci *Strep.* *pneumoniae, Klebsiella pneumoniae*

TOXIN-MEDIATED DISEASE

A number of important infections are mediated by bacterial exotoxins (Table 8.4) some of which act locally and some at a site distant from their production after blood-borne spread.

Table 8.4 Some toxin-mediated diseases.

Organism	Toxin	Disease
Gastrointestinal		
Staph. aureus	Enterotoxin	Vomiting, prostration
B. cereus	Enterotoxin	Vomiting, prostration or colitis
Cl. perfringens	Various enterotoxins	Diarrhoea, abdominal pain, blood in stools
Cl. difficile	Enterotoxin	Pseudomembranous colitis
Vibrio cholerae	Cholera toxin	Rice-water stools, dehydration, electrolyte imbalance
Esch. coli	Various toxins	Infantile, diarrhoea, travellers' diarrhoea
Esch. coli	Verotoxin	Acute haemorrhagic colitis, haemolytic uraemic syndrome
Central nervous system		
Cl. botulinum	Neurotoxin	Flaccid paralysis, headaches, vomiting
Cl. tetani	Neurotoxin	Spastic paralysis
Corynebacterium diphtheriae	Neurotoxin, cardiotoxin	Diphtheria
Skin		
Staph. aureus	Epidermolytic toxin	Toxin epidermal necrolysis
	TSST 1	Toxic shock syndrome
Strep. pyogenes	Erythrogenic toxin	Scarlet fever

BACTERIAL DISEASES CAUSED BY IMMUNOLOGICAL MECHANISMS

Various syndromes arise around 2 weeks after infection with certain strains of *Strep. pyogenes*. The pancarditis and arthritis of acute rheumatic fever is caused by antibody made to streptococcal proteins cross-reacting with specific proteins in the heart and joints. Post-streptococcal glomerulonephritis is due to immune complex deposition.

More tenous is the association between chronic carriage of certain coliforms in the gut with chronic arthritic disorders. For example, ankylosing spondylitis may be caused by antibodies reacting to common antigens on *Klebsiella* species from the gut and synovial cells in subjects who have a particular HLA constitution (HLAB27).

9 The Staphylococci

The staphylococci are Gram-positive cocci that are part of the normal flora of the skin. *Staphylococcus aureus* (coagulase-positive) can cause superficial or deep sepsis and coagulase-negative staphylococci can produce infections associated with prosthetic devices.

TRANSMISSION   Infection usually follows inoculation into a wound of staphylococci from the patient's own skin flora but in hospitals it is important to recognise that staphylocci may be spread by the hands of carers. *Staph. aureus* is carried in the nose of up to 40% of adults.

CLINICAL FEATURES   *Staph. aureus*: pustular skin disease, sycosis barbae, wound infection, 'scalded skin syndrome', toxic shock syndrome, diarrhoea and vomiting, pneumonia, acute osteomyelitis, septic arthritis, intravenous catheter infection, endocarditis.
   Coagulase-negative staphylococci: infection of intravascular catheters, prosthetic valves, grafts and joints, peritoneal catheters, and ventricular shunts, urinary tract infection.

THERAPY AND PROPHYLAXIS   Drainage of abscesses, removal of infected prosthetic devices and treatment with flucloxacillin, clindamycin or erythromycin, given with gentamicin or fusidic acid in severe infection. Multiply-resistant strains are treated with vancomycin or teicoplanin.

LABORATORY DIAGNOSIS   Staphylococci can be cultured from wound exudate, pus, body fluids, blood, or urine. Identification depends on morphology, the production of catalase and production or not of coagulase/clumping factor.

CLINICAL FEATURES

Staph. aureus causes pustular skin disease, for example folliculitis, boils, carbuncles, abscesses and sycosis barbae (folliculitis in bearded areas). It is the pathogen most commonly isolated from infected wounds following clean surgery. Inflammation is usually localised, spreading by direct extension, but systemic symptoms can result from release of extracellular products or organisms into the blood.

Exfoliative toxins are implicated in a more serious generalised infection, toxic epidermal necrolysis of 'scalded skin syndrome', which affects infants and young children. A staphylococcal exotoxin is responsible for toxic shock syndrome, in which a scarlatiniform rash and hypotension precedes abnormalities of various organ systems and widespread desquamation. Enterotoxin is produced by multiplication of

Staph. aureus in food and can produce diarrhoea and vomiting if ingested.

Staph. aureus causes pneumonia in infants, in immunosuppressed hospital patients and in adults following influenza infection. The latter frequently results in death in 1–2 days. *Staph. aureus* is the principal cause of acute osteomyelitis and septic arthritis. Septicaemia may originate from a wound infection, an intravenous catheter infection, osteomyelitis or endocarditis. Staphylococcal endocarditis in intravenous drug abusers usually affects the tricuspid valve and the patient can present with multiple staphylococcal abscesses in the lung. Endocarditis can be rapidly progressive and is associated with the development of abscesses in the valve ring and at distant sites. In elderly patients *Staph. aureus* tends to affect the aortic valve with rapidly progressive destruction of the valve.

Coagulase-negative staphylococci adhere to prosthetic materials and produce infection of intravascular catheters, prosthetic valves, grafts and joints, peritoneal catheters and ventricular shunts. Intravascular catheters are infected during insertion or manipulation and the likelihood of infection increases with time *in situ*. Coagulase-negative staphylococci cause one-third of cases of prosthetic valve endocarditis, particularly if it occurs within 6 months of operation. Fever can be the only sign but it is often accompanied by a new murmur or evidence of peripheral emboli. Late infection of joint prostheses presents with fever, pain or loosening and usually surgical removal. Infection of ventricular shunts occurs in up to 10% of cases and can present with wound infection and fever or non-specific symptoms, for example anorexia, weight loss, nausea and vomiting. Coagulase-negative staphylococci are a common cause of peritonitis in patients having peritoneal dialysis and, unlike *Staph. aureus*, arise from infection of the lumen of the catheter rather than the exit site.

Staph. epidermidis causes cystitis in elderly in-patients after instrumentation of the bladder and *Staph. saphrophyticus* is a common cause of symptomatic urinary infection in young healthy women.

THE BACTERIA

Staphylococci are non-motile spherical cells with a diameter of 0.5–1.5 μm. They aggregate in grape-like clusters (from which the name derives) and fresh cultures are Gram-positive. All staphylococci are catalase-positive and *Staph. aureus* also produces coagulase and usually has some capsular material, clinical isolates tending to have two of the eight possible serotypes. Coagulase-negative staphylococci comprise a range of species of which the most common human pathogen is *Staph. epidermidis*. This organism attaches to plastic and can use it as a food source. The mechanism of adherence is unclear but may involve the production of an extracellular slime. This substance contains 40% sugar and 27% protein, it can inhibit local host defences and it resists penetration by antibiotics.

EPIDEMIOLOGY

Staphylococci form part of the normal flora of the skin and mucous membranes. *Staph. epidermidis* is ubiquitous and *Staph. aureus* is found in 10–40% of adults in the anterior nares and in 10–20% at other sites. Between 60 and 90% of adults carry *Staph. aureus* in the nose at some time. Various coagulase-negative staphylococci inhabit the axillae and pubic areas, the scalp and the external auditory meatus. Most infections originate from the host's skin. However, staphylococci survive in fomites and are present in animal products, soil and water. Some nasal or perineal carriers of *Staph. aureus* shed large numbers of the organism as do some patients with skin infections or pneumonia. Wound infection can be caused by contact with staff carrying *Staph. aureus* on their hands. Colonisation of infants occurs largely by direct contact.

THERAPY AND PROPHYLAXIS

Abscesses usually require surgical drainage and this may be adequate treatment by itself. In the presence of cellulitis, a β-lactamase stable penicillin can be used, for example flucloxacillin 500 mg every 6 hours orally, with clindamycin or erythromycin as alternatives. Higher doses are needed in more severe infections. Penicillin should be used if the organism is sensitive. It is important to emphasise the need for careful hygiene, clean clothing and dressing of draining lesions to prevent reinfection. In toxic shock syndrome, vigorous fluid replacement is needed in addition to removal of the bacterial source.

Septicaemia or pneumonia due to *Staph. aureus* requires parenteral therapy with flucloxacillin 1–2 g every 4–6 hours, combined with gentamicin, or fusidic acid. Osteomyelitis and endocarditis need prolonged treatment (4–6 weeks). Clindamycin and fusidic acid have the best bone penetration. Methicillin-resistant *Staph. aureus* (MRSA) is not susceptible to flucloxacillin or cephalosporins but infections can be treated with vancomycin or teicoplanin. The patients must be source-isolated to prevent cross-infection. Infections of prostheses are usually caused by methicillin-resistant coagulase-negative staphylococci and respond most quickly to removal of the prosthesis. If this is impractical, the infection may be suppressed by vancomycin or teicoplanin combined with gentamicin or rifampicin. Peri-operative antibiotics, for example cefuroxime or flucloxacillin/gentamicin, are used in cardiac and orthopaedic surgery in an attempt to prevent prosthetic infections. Careful aseptic technique at insertion is of the greatest importance. Nasal carriage of MRSA can be eradicated by topical application of mupirocin.

LABORATORY DIAGNOSIS

Staphylococci can be isolated from wound swabs, or preferably pus, pleural or synovial fluid, sputum, urine and blood. The significance of isolates from non-sterile sites must be judged with respect to the clinical presentation. Particular care must be taken in the collection and examination of blood cultures to prevent both contamination of cultures with skin staphylococci and dismissal of significant isolates as contaminants. After overnight aerobic incubation on blood agar, staphylococci form smooth domed colonies 1–2 mm in diameter, some of which are beta-haemolytic. Only a proportion of strains of *Staph. aureus* develop the characteristic gold yellow colour.

The Gram stain may confirm recognition by colonial morphology. Catalase production indicates that the organism is a staphylococcus. Latex particles coated with human plasma agglutinate with clumping factor and staphylococcal protein A and are used to distinguish *Staph. aureus* from other staphylococci. Confirmation is provided by the coagulase test (formation of a clot after incubation of rabbit plasma with a broth culture) or by the production of DNAase (hydrolysis of nucleic acid in an agar medium). If necessary, species identification of coagulase-negative staphylococci can be made by commercially available biochemical test strips. The antibiotic susceptibility pattern, bacteriophage and plasmid typing can be useful in analysing the spread of staphylococci within the hospital. Production of exfoliative toxins and the toxin shock syndrome are associated with strains of certain bacteriophage groups.

Serological diagnosis is useful in some cases of deep infection, e.g. osteomyelitis, in which culture results are not available.

10 The Streptococci

The streptococci are Gram-positive cocci. Some will grow either aerobically or anaerobically, others are obligate anaerobes. There are many species widely distributed in man and other mammals. Streptococci may be harmless commensals, or give rise to localised infection (sometimes with immunologically determined complications locally or at a distance), or cause a fulminant septicaemic (and potentially fatal) infection.

TRANSMISSION   Oral, intestinal, genital tract or cutaneous carriage may lead to localised or widespread infection in the recipient by respiratory, intestinal, surgical or direct contact transmission, or on fomites. The incubation period is usually short, usually less than 48 hours. Some organisms have the capacity to spread rapidly in surgical or obstetric units.

CLINICAL FEATURES   *Strep. pneumoniae*: otitis media, sinusitis, acute bronchitis, pneumonia, meningitis.
Viridans streptococci: oral sepsis, bacteraemia/septicaemia, endocarditis, abscesses.
Strep. pyogenes: sore throat, tonsillitis, impetigo, wound infection, cellulitis, scarlet fever, puerperal fever.
Other beta-haemolytic streptococci: wound infection, neonatal infections, urinary tract infection.

COMPLICATIONS   These include septicaemia, abscess formation and immuno-logical sequelae.

THERAPY AND PROPHYLAXIS   They are generally antibiotic-sensitive organisms but penicillin is the treatment of choice. Abscesses require surgical drainage. Antibiotic prophylaxis of infective endocarditis is indicated in at-risk patients. A multivalent vaccine is available for prophylaxis against pneumococcal infections.

LABORATORY DIAGNOSIS   Streptococci can be cultured on simple media, from swabs of throat, skin lesions or wounds. Urine, blood, sputum of CSF samples may be cultured.

CLINICAL FEATURES

The streptococci may affect both sexes at all ages, causing a wide spectrum of disease.
Alpha-haemolytic streptococci (i.e. those that produce a green colour round colonies growing on blood agar) include *Strep. pneumoniae*, a respiratory tract commensal. It may cause respiratory tract infections, such as otitis media, sinusitis, acute bronchitis,

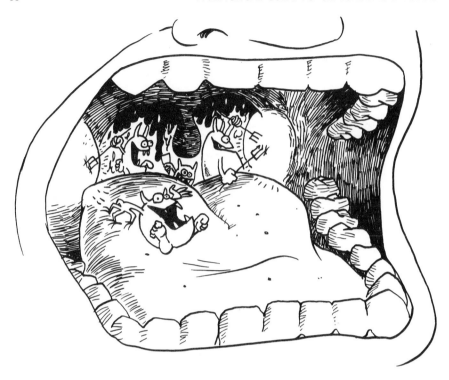

lobar pneumonia, bronchopneumonia, and septicaemia or meningitis.

The viridans group of streptococci includes many species, e.g. *Strep. durans, Strep. milleri, Strep. mitior, Strep. mutans, Strep. sanguis, Strep. salivarius.* They may cause dental caries, oral sepsis, bacteraemia, septicaemia, infective endocarditis. *Strep. milleri* may be associated with liver abscess or lung abscess.

Beta-haemolytic streptococci (lysis of blood around colonies produces completely clear zones in blood agar cultures). Lancefield Group A (*Strep. pyogenes*) may cause sore throat, tonsillitis, impetigo, wound sepsis, cellulitis, puerperal fever, scarlet fever (when the infecting organism produces erythrogenic toxin) erysipelas and necrotising fasciitis. Immunological sequelae may occur (see complications).

Lancefield Group B (*Strep. agalactiae*) may cause puerperal sepsis, neonatal septicaemia and meningitis, gynaecological wound infections, urinary tract infection (UTI).

Lancefield Group D (the enterococci, e.g. *Enterococcus faecalis*) cause UTI, gut-related sepsis, and infrequently cause infective endocarditis. Other Lancefield Groups (C,G, etc) cause sore throat, localised wound infection.

Anaerobic streptococci (*Peptococcus* species, *Peptostreptococcus* species) may cause abscesses, gut-associated sepsis, cellulitis, cerebral abscess, Melaney's synergistic gangrene (with other anaerobes, and staphylococci).

COMPLICATIONS

The most feared is a very rapid clinical course with overwhelming sepsis and a high case fatality. Rapid nosocomial spread giving rise to outbreaks of infection may also occur. Some strains of *Strep. pyogenes* give rise to immunological phenomena presenting as rheumatic fever (cross-reacting antibodies against streptococcal and cardiac antigens are thought to be responsible) and acute glomerulonephritis (an immune complex disease).

THE BACTERIA

The streptococci are Gram-positive cocci. They may be seen as single organisms, as pairs (diplococci), e.g. *Strep. pneumoniae*, or in chains, e.g. viridans streptococci, beta-haemolytic streptococci. (The name derives from the tendency to form chains.) *Strep. pneumoniae* may manifest capsules which may be seen by Gram staining or by the use of capsular stains.

The nature of haemolytic changes on blood agar cultures has been used for classification, although many strains do not behave typically. No haemolysis (gamma haemolysis) may occur round enterococcal colonies (*E. faecalis, E. bovis.*) Partial (alpha) haemolysis may be noted, with 'greening' due to the presence of altered haemoglobin round colonies of 'viridans' streptococci. Complete (beta) haemolysis may be seen with the clearing of blood pigment around colonies of *Strep. pyogenes*.

Further identification of alpha-haemolytic streptococci may be carried out. *Strep. pneumoniae* shows inhibition by ethylhydrocuprein hydrochloride (optochin) and is soluble in bile salts. Viridans streptococci are speciated by a range of biochemical tests.

Beta-haemolytic streptococci are classified into Lancefield Groups (e.g. A, B etc.) by extracting the polysaccharide antigens from the cells walls and reacting them with specific antisera. Lancefield Group A (*Strep. pyogenes*) strains may be further subdivided into types for epidemiological purposes by the extraction of M or T proteins from the cell wall and their identification by the use of specific antisera. Such typing is different from Lancefield grouping, and is used in the investigation of outbreaks of infection. The enterococci (e.g. *E. faecalis*) will grow on media containing bile (e.g. MacConkey agar) and will hydrolyse aesculin, producing a black colour in suitable media. Non-haemolytic streptococci may be identified biochemically and by the same tests described for alpha- and beta-haemolytic strains.

EPIDEMIOLOGY

Streptococci occur very widely. *Strep. pneumoniae* is very commonly present in the nasopharynx and most infections are endogenous in origin. *Strep. pyogenes* may be present in the saliva of 5–15% of well people and the enterococci and anaerobic streptococci are frequently present in the intestinal flora. Spread of respiratory organisms will be by inhalation of contaminated droplets of saliva. Hand-to-mouth transmission of respiratory and intestinal organisms is associated with poor hygiene. *Strep. pyogenes* may be spread among children with impetigo by direct physical contact. Impetigo, erysipelas and scarlatina have been associated with low social class, poverty, malnutrition and overcrowding. The immunological complications of *Strep. pyogenes* infection, e.g. rheumatic fever, glomerulonephritis and Henoch–Schönlein purpura, are related to strains with particular M or T types. The prevalence of such strains varies with time and in different places. Many of the complications were common in the UK formerly but are now rare. This might change again.

THERAPY AND PROPHYLAXIS

Strains of *Strep. pyogenes* are uniformly sensitive to penicillin and generally to erythromycin which may be used as an alternative in penicillin-hypersensitive patients. Lancefield Group B streptococci are of intermediate sensitivity to penicillin. At one time all *Strep. pneumoniae* were sensitive to penicillin. It is still the drug of choice but a few strains are resistant. Erythromycin, tetracyclines, chloramphenicol or vancomycin may be used as alternatives although none is absolutely dependable. *E. faecalis* is generally sensitive to ampicillin but not completely sensitive to penicillin or gentamicin. Most viridans streptococci are sensitive to penicillin.

Too much reliance should not be placed on antibiotics alone. Abscesses should be drained, attention must be paid to infection control and normal supportive care must not be neglected.

Patients with cardiac anomalies are considered to be at excess risk of contracting infective endocarditis from their oral flora (predominantly viridans streptococci) as a consequence of the transient bacteraemia associated with dental surgery. Such patients should be covered by antibiotic prophylaxis given shortly before the procedure. Antibiotics recommended include amoxycillin, erythromycin, gentamicin or vancomycin depending on the circumstances. Failure to arrange this may be considered negligent.

LABORATORY DIAGNOSIS

This is based on the culture of throat swabs, wound swabs, CSF, blood, urine etc. Most clinically important species of streptococci grow well on blood agar in 24 hours if incubated aerobically. Anaerobic streptococci may take several days to grow anaerobically. The addition of 5% carbon dioxide may enhance the growth of some streptococci. The colonies of *Strep. pneumoniae* may show concentric surface rings causing them to resemble draughtsmen. Species identification and antibiotic sensitivity testing are then performed.

Pneumococcal infection may be detected more speedily by antigen detection techniques thus saving time in, for example, suspected *Strep. pneumoniae* meningitis.

Treatment may have to anticipate culture results. The laboratory should report isolates of streptococci, particularly beta-haemolytic strains, as a matter of urgency (e.g. by telephone) lest delay compromises the recovery of patients with potentially overwhelming infection, or prevents infection control measures being taken as soon as possible.

11 The Corynebacteria

The corynebacteria are Gram-positive bacteria most of which grow aerobically or anaerobically, while some are obligate anaerobes. There are many species which are widely distributed in man and other mammals. They may be harmless commensals in the throat or on the skin or give rise to localised infection or occasionally cause septicaemia. One species, *Corynebacterium diphtheriae*, may cause localised throat infection with a risk of fatal respiratory obstruction in babies. Some strains of *C. diphtheriae* produce a powerful exotoxin capable of threatening the host's life by various mechanisms.

TRANSMISSION Respiratory strains are acquired by inhalation of respiratory droplets, skin and wound strains by direct contact or occasionally surgically or on fomites. Epidemic spread is possible in non-immune communities. The incubation period is 1–7 days.

CLINICAL FEATURES *C. diphtheriae* may give rise to pharyngitis, membrane formation in the larynx or local wound sepsis. With toxinogenic strains the exotoxin has effects on vascular tissues (shock), the nervous system (paralyses, inability to breathe), heart (myocarditis, shock) or kidneys (albuminuria).

Other species cause localised skin, throat or wound infections. *C. jeikeium* may cause a variety of serious systemic infections in hospitalised immunocompromised patients.

COMPLICATIONS These include septicaemia, toxaemia and wound sepsis.

THERAPY AND PROPHYLAXIS The corynebacteria are generally sensitive to penicillins and erythromycin and to a variety of other antibiotics. Management of diphtheria may include maintenance of airway, support of respiratory, cardiac and renal function, rehydration. Patients must be given diphtheria antitoxin *urgently*. Public health authorities must be notified at once.

Active immunisation against diphtheria is part of routine childhood protection programmes and has made diphtheria rare in developed countries.

LABORATORY DIAGNOSIS Corynebacteria can be cultured on simple media from swabs of throat, skin lesions or wounds.

CLINICAL FEATURES

The most important pathogen is *C. diphtheriae*, which may cause local lesions only. However, strains colonised by a specific non-lytic bacteriophage produce a potent exotoxin which results in non-immune hosts developing the disease, diphtheria. The

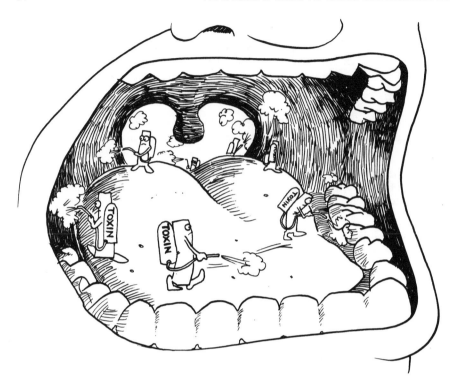

disease occurs in both sexes at any age. Outbreaks may occur and infection control is a major preoccupation. Patients usually present with fever, malaise and a sore throat. The acute inflammatory response leads to the formation of a pseudomembrane possibly involving the pharynx, larynx or tonsils.

Other features of diphtheria include shock due to toxic effects on the vascular system or to acute myocarditis causing heart failure, palatal and other paralysis including respiratory paralysis, and renal damage. The primary lesion is usually pharyngitis, but cases secondary to skin lesions, infected wounds or even vaginal lesions may occur. Death may occur as a consequence of acute respiratory obstruction or the cardiovascular changes.

C. haemolyticum, C. ulcerans, C. hofmanii, may all rarely be associated with localised infections of the throat or skin and soft tissues. *C. jeikeium* colonises hospitalised immunocompromised patients and is selected by exposure to antibiotics. It is increasingly being isolated from such patients as a cause of septicaemia, intravenous line infections, prosthesis-associated infections and a variety of skin and soft tissue infections.

COMPLICATIONS

The local complications are quinsy, wound sepsis, skin and soft tissue infections, acute respiratory tract obstruction. Systemically toxaemia, shock, myocarditis, cardiac arrhythmias, heart failure and palatal palsy, neurological sequelae, respiratory muscle paralyses, renal failure and albuminuria may occur.

THE BACTERIA

The corynebacteria are Gram-positive rods. Some, e.g. *C. diphtheriae*, are slightly curved and may be seen in groupings suggestive of Chinese pictograms. Metachromatic granules may be seen in *C. diphtheriae* or other corynebacteria exposed to Albert's stain.

Corynebacteria grow aerobically or anaerobically on blood agar and may be haemolytic. They may be selected from mixed culture by growth on culture media containing tellurite (e.g. Hoyle's medium) on which they give grey or black colonies. Such colonies may be characteristic ('daisy head') of particular types of *C. diphtheriae* of varying pathogenicity (*gravis, intermedius* or *mitis*) but such differentiation is nowadays used only, by the very few experts, for epidemiological studies. Serum–agar slopes (Loeffler's medium) are used to produce material for revealing characteristic microscopic morphology and in readiness for toxigenicity testing.

Corynebacteria may be speciated by differential fermentation tests. *C. diphtheriae* usually ferments glucose and maltose, and occasionally sucrose. In relation to the pathogenesis of human disease the vital information about *C. diphtheriae* strains is the production (or not) of the potent exotoxin which causes diphtheria in non-immune subjects. Toxin production is demonstrated *in vitro* by immuno-precipitation using an Elek plate.

EPIDEMIOLOGY

Corynebacteria occur as part of the normal human and animal commensal flora, particularly on the skin, in the upper respiratory tract, and in the vagina. *C. diphtheriae* is now a rare isolate in the predominantly immune populations of prosperous countries but may be more common elsewhere. Strains colonised by specific bacterial viruses (bacteriophages) produce exotoxin which manifests in man as the disease diphtheria. Droplet spread is the usual means of transmission, but direct contact may rarely transmit surgical or obstetric infection.

Many other corynebacteria may be isolated from cultures of clinical samples and must be differentiated from *C. diphtheriae*.

THERAPY AND PROPHYLAXIS

Local infection with corynebacteria including diphtheria of the larynx and oro-pharynx is treated with appropriate antibiotics such as erythromycin, various β-lactams or glycopeptides. Maintenance of a patent airway is important in diphtheria. This disease also presents as an exotoxaemia with general problems such as dehydration, confusion and fever, and specific problems such as myocarditis and cardiac arrhythmias, each of which should be managed appropriately. Large doses of *C. diphtheriae* antitoxin must be given at once when the condition is suspected. Infected patients must be isolated and managed by immune staff. Contacts must be traced actively and speedily, so cases must be notified at once by telephone to competent communicable disease control staff.

Prevention is highly effective and is achieved by the administration to children of diphtheria toxoid (inactivated toxin). A high proportion of the population should be kept immune as a matter of public health policy by achieving high immunisation rates in children.

LABORATORY DIAGNOSIS

This is based on the culture of throat swabs, nose swabs, wound swabs, blood etc. Corynebacteria usually grow readily and quickly. Species identification and antibiotic sensitivity testing are then carried out. Strains of *C. jeikeium* may require extended antibiotic sensitivity testing because of multiple resistances. Production of toxin by *C. diphtheriae* is demonstrated using an Elek plate in which toxins and the antitoxin added form a visible line of precipitate in the agar.

12 *Bacillus* Species

Bacillus spp. are Gram-positive or Gram-variable rod-shaped cells which grow best under aerobic conditions. They can form endospores and are widely distributed in nature, particularly in the soil. The majority of species are not pathogenic but *Bacillus anthracis* and *Bacillus cereus* are important causes of disease.

TRANSMISSION   Man is infected with *B. anthracis* by contact with animals or animal products. Pulmonary disease is acquired by inhalation of an aerosol of spores. Most anthrax infections are cutaneous with an incubation period of 2–3 days. Pulmonary and gastrointestinal disease are rare but have a high mortality. *B. cereus* causes vomiting or diarrhoea 5–8 hours after ingestion of contaminated rice or other foodstuffs. It can cause ocular infections, pneumonia, endocarditis and bacteraemia.

CLINICAL FEATURES   Cutaneous anthrax is characterised by a painless ulcer surrounded by vesicles and oedema. Pulmonary disease results in dyspnoea, hypoxia and death. Abdominal pain, bleeding and ascites indicate gastrointestinal disease.

   B. cereus can cause vomiting or diarrhoea, an illness with a rapid onset but which resolves within 24 hours. Severe ocular infections and endocarditis occur in drug abusers. Bacteraemia is often related to infection of intravascular devices.

COMPLICATIONS   These are septicaemia and meningitis.

THERAPY AND PROPHYLAXIS   Anthrax should be treated with benzylpenicillin. Patients must be isolated and the disease notified. Individuals and animals at risk should be vaccinated. *B. cereus* food poisoning requires only supportive measures, although the rare invasive disease requires treatment with vancomycin or clindamycin.

LABORATORY DIAGNOSIS   *B. anthracis* can be cultured from ulcer fluid or blood. Material known to be infected should be manipulated in a safety cabinet. *B. cereus* is a common laboratory contaminant but for it to be judged a significant pathogen repeated isolation or its presence in large numbers, e.g. in food, is required.

CLINICAL FEATURES

B. anthracis

In the cutaneous form of anthrax, a papule appears 2–3 days after infection. This ulcerates, becomes surrounded with vesicles and then blackens forming an eschar. In most cases, the lesion heals in 2–6 weeks but, if untreated, a fifth of patients develop headache, fever and nausea with septicaemia, resulting in shock and death. In intestinal disease, the eschar forms in the ileum or caecum causing marked local

oedema. Abdominal pain, fever, vomiting and bloody diarrhoea follow and in some cases, the patient rapidly becomes shocked and dies. The pulmonary form is usually fatal, although the onset is gradual. It presents as a virus-like respiratory infection for 2–3 days and then progresses to dyspnoea, tachycardia and cyanosis. Fever develops with rapid onset of coma.

B. cereus

When ingested in large numbers, profuse diarrhoea and abdominal pain develop 8 hours after ingestion but symptoms resolve within 24 hours. In outbreaks associated with fried rice, nausea and vomiting are the main complaints occurring within 5 hours of ingestion. Enterotoxins are responsible for both the diarrhoeal and emetic forms. B. cereus is a common cause of post-traumatic endophthalmitis, particularly in rural areas, and is characterised by rapid tissue destruction and formation of a ring abscess. Pneumonia and necrotising fasciitis in the immunocompromised patient have been reported and B. cereus can cause endocarditis and osteomyelitis in drug abusers. Bacteraemia with Bacillus species is often associated with infection of intravascular catheters.

COMPLICATIONS

Septicaemia is frequently fatal and can accompany any form of anthrax but is common in pulmonary disease. Meningitis can develop after cutaneous anthrax and is usually fatal. Other *Bacillus* species can cause meningitis following otitis or urinary tract infection or spinal anaesthesia.

THE BACTERIA

Bacillus species are rod-shaped organisms which vary greatly in size and occur singly or in chains. The cells stain Gram-positive but this is lost in older cultures. Endospores develop in air and these are particularly important in the persistence and spread of the organism. With the exception of *B. anthracis*, the organisms are motile. Capsules can be formed under appropriate conditions.

Bacillus species are aerobic but *B. cereus* and *B. anthracis* can grow anaerobically. Growth is optimal between 25°C and 37°C but certain species, e.g. *B. stearothermophilus*, can grow at 55°C or more. Spores resist heating at 70°C for 10 minutes which is useful in testing autoclaves. The formation of spores in air and the production of catalase distinguish *Bacillus* species from the clostridia. Selective media are available for the isolation of *B. anthracis* and *B. cereus*. Species identification can be made by the growth conditions, production of lecithinase, biochemical reactions and appearance of the spores.

EPIDEMIOLOGY

Anthrax is principally a disease of cattle, sheep, goats and other herbivores. Infection in man occurs following direct contact with infected animals, e.g. farmers, butchers or veterinary surgeons, or from contact with spores present in hides, wool or animal products. Pulmonary disease only occurs when an aerosol of spores is inhaled in processing of animal products, e.g. goat hair. *B. anthracis* has a reservoir in the soil but specific and poorly understood conditions are needed for multiplication sufficient to produce disease in animals. Human disease is rare in developed countries because of vaccination of animals and workers, improved sterilisation and hygiene but persists in Africa, India and the Middle East.

Other *Bacillus* species are widespread in soil, water, vegetables and as normal human gut flora. Food poisoning due to *B. cereus* has been associated with the use of spices containing large numbers of organisms and boiled or fried rice which has been prepared in bulk and kept at ambient temperatures before reheating. Immunocompromised patients are usually infected from their own flora and drug abusers by their injection equipment or the heroin itself.

THERAPY AND PROPHYLAXIS

The treatment of choice of cutaneous anthrax is benzylpenicillin for 2–4 days followed by oral penicillin for a total of 10 days. Pulmonary and gastrointestinal diseases are treated with high doses of penicillin but the success rate is low. *B. anthracis* is sensitive to tetracyclines, chloramphenicol, gentamicin and erythromycin which may be used in patients with a history of allergy to penicillin. The disease is notifiable. Patients should be source isolated and transferred to a regional infectious disease unit. Pulmonary disease is particularly infectious. Infected animals must be burnt or buried, and animal vaccines used in infected areas. Persons likely to be exposed to the spores should be vaccinated.

Food poisoning due to *B. cereus* requires symptomatic treatment and fluid replacement but not antibiotics. Early refrigeration of boiled rice helps prevent germination of spores that have survived cooking. Ophthalmitis requires aggressive treatment, i.e. vitrectomy and administration of clindamycin and gentamicin systemically and gentamicin into the vitreous. Endocarditis has been treated with vancomycin. Intravascular catheters should be removed if infected and the patient is symptomatic. Immunocompromised patients should be given vancomycin or clindamycin with an aminoglycoside.

LABORATORY DIAGNOSIS

B. anthracis can be seen in a Gram stain of vesicle fluid and cultured from the fluid and blood. Its lack of motility, presence of a capsule, growth in long chains and the lack of turbidity in broth distinguish it from other species. It is a hazard group 3 pathogen and procedures with infective material must take place in a microbiological safety cabinet.

B. cereus is a common laboratory contaminant and widely distributed in the environment. To implicate it in a case of food poisoning requires isolation of the same serotype from the food and the faeces or isolation of more than 100 000 colony forming units of the organism per gram of food and/or faeces. Diagnosis of pneumonia depends on isolation from a deep specimen, e.g. needle aspiration. Repeated isolation from blood cultures is necessary for diagnosis of endocarditis. Isolation from blood and catheter tip confirms bacteraemia due to an intravascular device.

13 *Listeria*

The only human pathogen in this genus is *Listeria monocytogenes* which is a slender Gram-positive bacillus found widely in the environment and capable of causing infections in numerous different animal species. In humans it causes septicaemia, meningitis and abortion.

TRANSMISSION   The organism has a wide natural reservoir in both the environment and many different animals. Infection occurs after contact with infected animals, ingestion of contaminated food or transplacentally.

CLINICAL FEATURES   The infection can be asymptomatic or present with a flu-like illness, septicaemia or meningoencephalitis. A severe disseminated infection can occur in neonates following transplacental infection.

THERAPY AND PROPHYLAXIS   Because of the extensive natural reservoir the control of listeriosis depends on prevention of transmission to humans. Treatment is usually with ampicillin and gentamicin for 2–6 weeks.

LABORATORY DIAGNOSIS   *Listeria* can be isolated from blood, CSF and placenta. The colonies are beta-haemolytic and the bacteria have a characteristic tumbling motility. Identification is by biochemical profile.

CLINICAL FEATURES

Neonatal Listeriosis

Perinatal listeriosis is acquired transplancentally. Infection in the mother may be asymptomatic or present as a mild, febrile, influenza-like illness with pharyngitis, lymphadenopathy and malaise and can lead to abortion. Alternatively delivery may occur with the neonates suffering from a septicaemic illness, characterised pathologically by widespread granulomatous lesions in many organs (granulomatosis infantiseptica). The delivery is often precipitous, may be premature and the amniotic fluid is stained with green meconium.

Infections sometimes occur within a few days of birth and the neonate develops meningoencephalitis. The source of the organism in this case is probably the environment or another infected neonate.

Listeriosis in the Adult

Listeriosis in the immunocompromised host (renal or bone marrow transplant recipients, AIDS cases) is characterised by a pyrexia of unknown origin or a meningoencephalitis.

Listeriosis may develop after occupational exposure in veterinarians and farm workers or in otherwise healthy individuals. The illness can present as a mild, flu-like illness, similar to that occurring in pregnant women, a septicaemia or meninogoencephalitis similar to that in immunocompromised patients or rarely as focal skin lesions.

THE BACTERIA

Listeria are aerobic, non-spore forming, Gram-positive coccobacilli that exhibit pleomorphism. They are $0.2–0.4\,\mu m \times 0.2–0.5\,\mu m$ in size but may be confused microscopically with cocci or diphtheroids. They are motile and a characteristic form of tumbling motility may be seen. There are currently eight species, but only *L. monocytogenes* is pathogenic. The bacteria grow readily on blood agar over a wide temperature range ($4–37°C$). *L. monocytogenes* is beta-haemolytic. It can be divided into a number of serological groups based upon somatic or O antigens 1 to 15 and flagella or H antigens a to e.

EPIDEMIOLOGY

Listeria has a very wide natural reservoir. It is found in soil, water, sewage, pigs, sheep, goats, cattle, cats, dogs, rodents, fish, birds and insects. Human infection occurs after direct contact with live infected animals; ingestion of contaminated food (food-source animals, vegetables, milk or soft cheeses) and transplacentally. Over 90% of human infection is caused by serotypes 4b, 1/2a and 1/2b. The annual infection rate is currently 2.3/million population in the UK and the majority of infections occur in patients who have an underlying illness. The second major group of individuals to be infected are pregnant women. A small number of cases occur in patients who have no obvious risk factor. Faecal carriage in the general poplulation is low, but may be up to 30% of symptomatic cases. Faecal carriage does not seem to persist for more than 1 month. Because of the ubiquitous nature of *Listeria* control of human infection depends on identification and slaughter of infected food source animals, pasteurisation of milk, and adequate food preparation and storage. Some foods (e.g. soft cheese) should be avoided during pregnancy.

THERAPY

Most infections are treated with β-lactam antibiotics (ampicillin, azlocillin, benzylpenicillin) often combined with gentamicin. The length of treatment is 2 weeks, but in immunocompromised patients 4 or 6 weeks may be preferable. Treatment failures have occurred with this regimen and alternative antibiotics that have been used successfully are tetracycline, chloramphenicol, erythromycin (combined with tetracycline or alone) and sulphonamides.

LABORATORY DIAGNOSIS

Listeria can be isolated from blood cultures, cerebrospinal fluid (CSF), the placenta and focal lesions in neonates with disseminated infection.

The specimen is incubated on blood agar and the colonies are surrounded by a narrow zone of beta-haemolysis. Because of the scanty numbers that may be found in CSF and the pleomorphism of *Listeria* they can be confused with enterococci or diphtheroids and apparent contaminants must be investigated fully in patients in which listeriosis is suspected. A motility test will reveal the characteristic tumbling motility of the organism. A full biochemical profile should be determined to identify the organism.

When attempting to isolate *Listeria* from food or faeces, the specimen should be inoculated into a broth, kept at 4°C and subcultured at weekly intervals.

14 The Clostridia

The clostridia are Gram-positive rods. They are generally motile, and produce spores which are usually heat resistant. The group is anaerobic, although a few strains will tolerate microaerophilic conditions. Their main habitat is the soil, but they are also widespread in man and other animals, forming part of the normal bowel flora. Clostridia cause a variety of clinical syndromes ranging from wound infection and occasionally bacteraemia, to toxin-mediated gas gangrene, enteritis, botulism and tetanus.

TRANSMISSION They are ingested in many foodstuffs, both cooked and uncooked. They may contaminate skin or be inoculated via puncture wounds or at operation. Spores may survive chemical and heat disinfection and be present on instruments used for sterile procedures.

CLINICAL FEATURES The clinical features depend on the species of *Clostridium* and the site of infection. Wound infection may occur with *Cl. perfringens* and *Cl. tetani*, the former producing destructive lesions in the neighbouring tissue and the latter causing a toxin-mediated disease characterised by muscle spasm. *Cl. botulinum* also causes a toxin-mediated disease with central nervous system manifestations. *Cl. perfringens* may cause a typical food-poisoning illness after eating contaminated food. Finally, pseudomembranous colitis can occur in patients receiving broad spectrum antibiotics due to the overgrowth of *Cl. difficile* in the gut.

THERAPY AND PROPHYLAXIS Clostridia are sensitive to penicillin, metronidazole and vancomycin. Surgical debridement and hyberbaric oxygen therapy are indicated in the management of clostridial gas gangrene. There is a mass immunisation programme for the prevention of tetanus. Botulism is prevented by safe bottling and canning procedures and good food handling practices prevent food poisoning.

LABORATORY DIAGNOSIS A Gram stain of wound specimens may be helpful and these and blood specimens should be cultured under anaerobic conditions. Some organisms are difficult to isolate; therefore toxin detection is used for diagnosis.

CLINICAL FEATURES

Wound Infection, Gas Gangrene

Traumatic and operation wounds, particularly those involving the lower body, are frequently contaminated by clostridia. *Cl. perfringens* is the most common species involved. Anaerobic conditions are required for the development of severe sepsis,

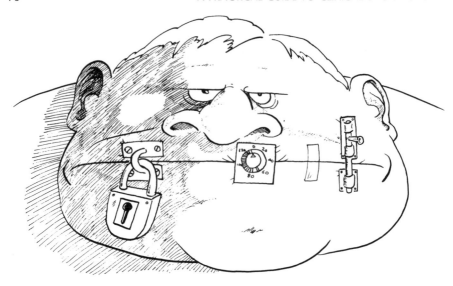

such as may be found in deep puncture wounds, crush injuries and after lower limb amputation. Under these conditions, gas is produced in the tissues, which may be detected by palpation or X-rays. Gas gangrene is characterised by profound toxaemia due to exotoxins produced by the infecting bacteria. Onset occurs within days of injury. Increasing pain, oedema, a dusky hue to the skin and a serous discharge are early signs. Gas may not be present at this stage. Constitutional symptoms and signs include sweating, fever, tachycardia and agitation. At operation, affected muscle does not contract when stimulated, and varies in colour from dark red to purple.

Enteritis

Clostridial food poisoning starts 6–12 hours after the consumption of food contaminated by *Cl. perfringens* spores and is characterised by cramp-like abdominal pain with profuse diarrhoea. Vomiting and pyrexia is unusual. The condition is usually self-limiting within 12–24 hours.

Pseudomembranous Colitis (PMC)

Superinfection with *Cl. difficile* related to current or recent antimicrobial chemotherapy produces a wide range of symptoms from mild, self-limiting diarrhoea to severe toxic colitis. Diarrhoea associated with antimicrobial chemotherapy is very common, and the presence of *Cl. difficile* or its toxin does not necessarily mean that the patient has PMC. In the latter condition pseudomembranous plaques are visible on the mucosal wall in the majority of cases on sigmoidoscopy.

Botulism

This is a toxin-mediated disease following ingestion of food contaminated by *Cl. botulinum* spores. Symptoms develop 12–72 hours after ingestion, and vary from mild weakness to collapse and rapid death. Early symptoms include nausea and vomiting, and dryness of the throat and mouth. Neurological symptoms may predominate with diplopia and dizziness, progressing to difficulty with speech and swallowing, and muscular weakness. Respiratory embarrassment may result. Symptoms may result from contamination of wounds with botulinum spores. Infant botulism can occur as a consequence of gut infection early in life.

Tetanus

Tetanus may result from the inoculation of spores of *Cl. tetani* into a wound. Onset is usually within 14 days of injury. The initiating wound may be minor or unknown. Onset is usually within 14 days of injury. The initiating wound may be minor or unknown. Dominant features of the condition are rigidity and spasm. Trismus or 'lock-jaw' results from stiffness of the jaw muscles, whilst stiffness of the facial muscles results in the sneer-like 'risus sardonicus'. Increasing involvement of other muscle groups leads to arching of the back and neck (opisthotonus). Spasms, involving several muscle groups, may be superimposed on rigidity. Splinting of the thorax may affect respiration, with frequent spasms leading to cyanosis and respiratory failure.

COMPLICATIONS

Uncomplicated contamination of wounds is of little consequence, and enteritis is usually self-limiting. Gas gangrene is a life-threatening infection requiring urgent surgical intervention in addition to supportive measures. PMC responds to prompt chemotherapy. Botulism and tetanus require a high level of supportive care, as the progression of the condition must be monitored since the more severe the symptoms, the greater the danger of death.

THE BACTERIA

Clostridia are large Gram-positive rods. Spores may be visible, although this is unusual for *Cl. perfringens* in samples from tissue. *Cl. tetani* spores are terminal, giving a drumstick appearance to the organism. Other species show subterminal spores. Most species are motile (*Cl. perfringens* is an important exception) and will swarm over a blood agar plate. Haemolysis is variable.

Carbohydrate fermentation reactions and lecithinase or lipase production are used to identify the species. The use of egg-yolk medium to demonstrate lecithinase production and its inhibition by specific antitoxin is the basis of the Nagler reaction.

EPIDEMIOLOGY

Clostridia are primarily soil commensals, where they lead a saprophytic existence, entering the food chain via ingested vegetables or contaminated meat. They colonise the bowel of man and animals. Wound infection (except for tetanus) usually follows auto inoculation, and remains localised.

Inadequate or partial cooking of meat or meat containing dishes e.g. stews, may lead to preservation and proliferation of *Cl. perfringens*.

Home canning with failure to heat the food adequately is particularly, but not exclusively, associated with human botulism.

Tetanus spores present in soil and dirt are likely to be inoculated following any traumatic injury.

THERAPY AND PROPHYLAXIS

Clostridia are sensitive to penicillin and metronidazole. Alternative antimicrobials include macrolides, lincosamines and tetracyclines, but acquired resistance is increasing. They are resistant to aminoglycosides. Vancomycin is used for the oral therapy of PMC.

Hyperbaric oxygen is indicated for the treatment of clostridial gas gangrene along with antitoxin, but surgical debridement and penicillin are most important.

Clostridial food poisoning and botulism are best prevented by education as to the safe preparation and preservation of food.

Mass active immunisation against tetanus should be encouraged. Specific tetanus antiglobulin should be given followed by a booster dose of vaccine after dirty wounds, especially if they are more than 6 hours old, If the patient has not been immunised previously, or the status is unknown, a full course of vaccine should be given.

LABORATORY DIAGNOSIS

Gram stain of wound exudate may reveal stumpy non-sporing Gram-positive rods. Anaerobic culture of wound specimens and blood should also be performed to confirm the diagnosis. Clostridia can be isolated on blood agar incubated anaerobically or in anaerobic broth media (e.g. cooked meat).

Laboratory tests are of little help for the diagnosis of botulism and tetanus. Botulism can be confirmed by demonstration of the toxin by intraperitoneal inoculation of serum or urine into mice.

The specific treatment of toxin-mediated clostridial diseases is frequently necessary on clinical suspicion alone.

15   *Actinomyces/Nocardia*

Amongst the group of Gram-positive branching rods which used to be thought of as fungi there are many genera. Only two are of interest to clinicans: *Actinomyces* (anaerobic) and *Nocardia* (aerobic).

Actinomyces

Actinomycetes are strictly anaerobic non-motile, branching, rod-shaped bacteria 1 μm in diameter. The colonies are smooth fast-growing or slow-growing, with thin spreading filamentous spider colonies and are hard looking after 7 days, like a 'molar tooth'.

Actinomycetes are normal commensal flora of the mouth and gut. They cause abscesses in the jaw and in the region of the caecum. The best known species is *Actinomyces israelii*. However, several others are involved but they are difficult to differentiate from one another.

CLINICAL FEATURES

Actinomyces israelii causes deep-seated abscesses, for example arising in dental sockets, leading to chronic osteomyelitis and soft tissue infection.

Infection arises from the normal commensal flora. Abscess formation occurs spontaneously or after trauma. The trauma includes dental work but orofacial disease is rare in people who have good dentition. The abscesses are cold, chronic and will eventually break down and discharge from multiple sites. The pus characteristically contains 'sulphur granules' which are small (<1 mm), hard white-yellow granules which represent colonies of the organism surrounded by inflammatory cells.

Spontaneous, chronic, deep-seated abscesses in the abdomen may arise and are often in the right iliac fossa, related to the appendix. These may be silent and then present with masses which form sinuses in the anterior abdominal wall. Other rare conditions include thoraco-pulmonary actinomycosis which probably follows inhalation of mouth contents or even a tooth. A local pneumonitis with abscess formation will occur and eventually erode through pleura and chest wall. Endometritis is associated with long-term use of some old-fashioned intrauterine contraceptive devices. Beware reports of *Actinomyces* in cervical smears since lactobacilli and

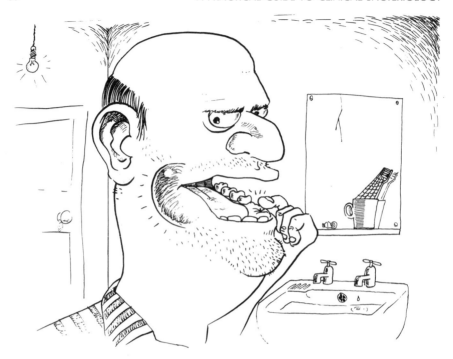

anaerobic coryneforms, which are normal vaginal flora, look similar. Immunofluorescent antibodies against *Actinomyces* will differentiate these genera on smears but this test is not routinely done.

Disseminated actinomycosis is rare but may follow any abscess. Sometimes actinomycetes are isolated from superficial or deep abscesses of the lower limb and such infection may be diffuse.

EPIDEMIOLOGY

Actinomyces abscesses are associated with poor dentition (some species specifically with plaque formation), intrauterine devices and caecal masses. The organisms are found in the normal flora in the mouth, gut and vagina (low numbers).

THERAPY

Surgical debridement is generally required. Diagnostic biopsies should be taken to differentiate these lesions from tumours or, in the gut, Crohn's disease. Following diagnosis or debridement, penicillin should be given.

LABORATORY DIAGNOSIS

Histological examination of the resected masses and abscesses will reveal characteristic colonies but only when stained with Gram (positive) or Grocot's silver stains. The organism is slow to grow, characteristically being inoculated onto blood agar (with neomycin to stop the growth of coliforms), and incubated under strict anaerobic conditions for at least 5 days. The culture can be enriched by inoculation into Robertson's cooked meat medium. The organism produces volatile fatty acids in liquid culture detectable by gas–liquid chromatography which will differentiate genera.

Nocardia

Nocardia are aerobic often branching, Gram-positive, acid-fast rods which cause abscesses in immunocompromised patients. The most important species is *Nocardia asteroides*. Several other similar species and genera are involved in mycetomas.

CLINICAL FEATURES

Infection is rare in healthy individuals. The commonest infection is a subacute respiratory infection with consolidation, abscess formation and pleuritic involvement. *Nocardia* may also infect skin and soft tissue causing chronic abscesses. There may be secondary bloodstream spread to brain, kidney, bone and elsewhere.

The infection is chronic and in many ways behaves like reactivation tuberculosis in immunosuppressed patients, though the lesions of nocardiasis tend to be in the lower lobes. Disseminated nocardiasis has been described in solid organ transplant recipients occasionally in outbreaks, and more recently in AIDS.

EPIDEMIOLOGY

The organisms are found in soil and dust (including in hospitals) and infection is acquired from the environment usually by inhalation. Person-to-person spread is rarely proven, though outbreaks occasionally occur in solid organ transplant recipients.

THERAPY

Though sensitive to many antibiotics *in vitro*, clinical experience shows sulphonamides to be the treatment of choice. Addition of amikacin (but not other aminoglycosides) is helpful.

LABORATORY DIAGNOSIS

Specimens may contain scanty small white-yellow 'sulphur' granules as in actinomycosis. The organisms stain Gram-positive and with Grocott's silver stain and are also partly acid-fast if decolorised with weak acid (use modified Ziehl–Neelsen or auramine stains). The organism grows on blood agar over several days and at high temperature which may be helpful to differentiate from other species.

16 *Treponema* Species

Treponema species are slender, tightly coiled rods 5–20 µm in length which are difficult or impossible to culture *in vitro*. *Treponema pallidum* comprises three very closely related subspecies, *pallidum, pertenue* and *endemicum*, which are respectively responsible for venereal syphilis, yaws, and endemic syphilis or bejel. *Treponema carateum* causes pinta, a chronic skin disease in Latin America.

TRANSMISSION   Syphilis is spread by sexual intercourse, the primary chancre appearing at the site of inoculation after 3 weeks. Secondary syphilis develops 2–8 weeks later, marked by a characteristic rash and widespread lymphadenopathy. Yaws and pinta affect children and are spread by contact, papules developing 1–5 weeks after infection. Bejel is spread by contact in warm climates where hygiene is poor but is not usually recognised until the secondary disease develops.

Oral treponemes are part of the normal mouth flora but proliferate in subgingival plaque during or before the development of periodontal disease.

CLINICAL FEATURES   The primary chancre of venereal syphilis is a painless papule which heals in 3–6 weeks. Secondary syphilis is associated with a maculopapular rash, condylomata lata, fever, weight loss and generalised arthralgia. Relapses occur for up to 4 years. Yaws presents as papules on the lower limbs which recur for 5 years and are associated with osteitis and deformation of the long bones in children. The papules of pinta are replaced by secondary lesions which heal with discoloration of the skin. Treponemes are associated with necrotic infections of the pharynx and gingival margins.

COMPLICATIONS   Neurosyphilis presents 5–30 years after infection with personality change, paresis and tabes dorsalis. Cardiovascular syphilis give rise to an ascending aortic aneurysm and incompetence of the valve. Congenital infection can occur. Gummata form in the skin and bones in both syphilis and yaws.

THERAPY AND PROPHYLAXIS   Penicillin is the treatment of choice. There is no vaccine for syphilis and sexual contacts should be treated. Yaws and pinta can be prevented by treating contacts and latent infections.

LABORATORY DIAGNOSIS   Treponemes can be seen by dark-ground microscopy of fluid expressed from skin lesions. Serological examination by the rapid plasma reagin test, *T. pallidum* haemagglutination assay and fluorescent treponemal antibody test provide the diagnosis in all except the earliest cases.

CLINICAL FEATURES

In venereal syphilis, a painless papule, the primary chancre, appears an average of 21 days after infection, accompanied by regional lymphadenopathy. It erodes and heals in 3–6 weeks.

Secondary syphilis starts 2–8 weeks later and is characterised by a maculopapular or pustular rash that begins on the trunk but can spread to any part, including the palms. In warm moist areas, wart-like growths known as condylomata lata can form. Fever, pharyngitis, weight loss, arthralgias and generalised lymphadenopathy occur. Headache, meningism, proteinuria and hepatitis develop as organs are invaded. Relapses may occur for up to 4 years before the disease becomes latent.

Yaws presents as papules, usually on the legs, that appear 3–5 weeks after infection, erode and heal within 6 months. The rash relapses repeatedly for the first 5 years and osteitis develops which can affect long bones (sabre tibia). Pinta presents 1–3 weeks after infection with chronic pruritic cutaneous papules. The small secondary lesions develop at the same sites after 3–12 months and recur for 10 years, healing with discoloration. Bejel is usually seen first as the secondary disease with mucous patches, condylomata lata and lymphadenopathy. Late lesions, gummata of the skin and bones, are a common presentation.

Vincent's angina is a rare ulcerative necrotic infection of the pharynx caused by anaerobic bacteria and spirochaetes. There is a purulent exudate and the breath is foul-smelling. Treponemes are common in periodontal pockets and may be responsible for periodontal disease.

COMPLICATIONS

Late syphilis can involve any organ and develops in one-third of untreated patients. Neurosyphilis is often asymptomatic but from 5 to 30 years after infection can result in change of personality with delusions, general paresis, memory loss, slurred speech and tabes dorsalis: ataxic gait, loss of proprioception, paraesthesia, lightning pains, incontinence, a small pupil unreactive to light (Argyll Robertson) and rarely, degenerative joint disease (Charcot's joints). In cardiovascular syphilis, obliteration of the small vessels supplying the wall of the aorta results in the formation of an aneurysm, usually in the ascending portion, and aortic valve incompetence. In skin or bone, nodules can form which erode into chronic ulcers (gummata).

An untreated infection in a pregnant woman results in congenital disease characterised in the infant by a desquamative rash, splenomegaly, anaemia, jaundice, deafness, peg-shaped teeth (Hutchinson's teeth) and osteochondritis giving a saddle nose and sabre tibia.

In the late stage of yaws, gummatous lesions of the bone occur with nodules and ulcers in the skin. In pinta, depigmented lesions are visible on the extremities.

THE BACTERIA

Treponema species are small Gram-positive helical rods 0.1–0.4 μm in diameter and 5–20 μm long with 6–14 spirals. They have cytoplasmic filaments running the length of the cell and three to four endoflagella at each end, which are needed for their rotational and forward motion. They are difficult to stain and dark-ground or phase-contrast microscopy are commonly used. Human pathogens are thought to be microaerophilic but have not been grown reliably *in vitro*. They can be grown in the rabbit testis but the doubling time is at least 30 hours. Testing of susceptibility is therefore limited but firm evidence of resistance to penicillin is lacking. Numerous polypeptide antigens can be identified by gel electrophoresis and immunoblotting. Serological techniques suggest that the differences between *T. pallidum* subsp. *pallidum* and subsp. *pertenue* and *T. caraetum* are small and they are indistinguishable by morphology or DNA hybridisation. However, the nature of the surface of treponema *in vivo* is uncertain as changes occur rapidly during incubation *in vitro*.

Some oral treponemes, for example *T. denticola* and *T. vincentii*, can be cultivated *in vitro* under anaerobic conditions in the presence of fatty acids. However, they probably represent only a few of the species present in the mouth and may not be the most pathogenically important.

EPIDEMIOLOGY

In syphilis, infection occurs through abrasions in the skin or mucosa. Survival outside the body is short so transmission is usually venereal and often within the first 2 years of the disease. Infection can occur by kissing, blood transfusion or by passage through the placenta. Bejel affects children in areas of poor hygiene and living standards. It is spread by contact and by common food and drink utensils.

After a period of slow multiplication, the primary chancre, heavily loaded with treponemes, is produced at the site of inoculation. Secondary lesions also contain large numbers of organisms. In late syphilis, few treponemes remain, most of the clinical effects being related to the host response. The prevalence of syphilis in the USA is increasing, possibly in association with human immunodeficiency virus infection.

Humoral immunity develops soon after infection but antibodies are ineffective in killing treponemes. However, they are useful for diagnosis and reinfection is rare. Cell-mediated immunity develops later and has some protective effect.

Yaws and pinta affect children and are spread by contact between infectious skin lesions in patients and the broken skin of susceptibles. Yaws is present in Africa, South America and Southeast Asia but pinta is limited to Central and South America.

Spirochaetes are part of the normal oral flora, particularly in the gingival crevice, forming 1–3% of dental plaque. In periodontal disease and gingivitis, they constitute 10–90% of subgingival plaque. However, their presence does not always correlate with disease. *Treponema denticola* produces hydrolytic and fibrinolytic enzymes and ammonia that may be important in causing periodontitis.

THERAPY AND PROPHYLAXIS

Treponemes are sensitive to penicillin. Primary and secondary syphilis should be treated with high-dose regimens. Procaine penicillin is given by daily intramuscular injection for 10 days with probenecid or benzathine penicillin is given in 2–3 weekly doses. A 3 week course of procaine penicillin is used in late syphilis. Resistance to penicillin has not been proven but treatment failures do occur. Fever, headache and worsening of the rash (Jarisch–Herxheimer reaction) can occur during treatment in early disease and steroids may be needed to prevent severe reactions in late disease. Patients treated for neurosyphilis should be followed by repeated serological testing for 5 years.

A single injection of benzathine penicillin is curative in yaws, pinta and bejel but chloramphenical and tetracycline have been used in courses of 10–14 days. Lesions heal in 1–2 weeks in yaws but take up to 12 months in pinta. Transmission can be prevented by treating contacts and latent cases. Improvements in living conditions and hygiene are effective in eradicating bejel.

Penicillin is the treatment of choice for Vincent's angina. Periodontitis can be prevented by plaque control and scaling but tetracyclines may be needed in treatment.

LABORATORY DIAGNOSIS

Treponemes can be seen by dark-ground or phase contrast microscopy in fluid expressed from the surface of the cutaneous lesions of syphilis, yaws, pinta or bejel. Direct fluorescent antibody staining of a fixed smear is useful for confirmation, and to demonstrate organisms which are immobile or from oral lesions or tissue samples.

Serological tests are the principal method of diagnosis of syphilis, although early primary disease may require repeated tests. The rapid plasma reagin (RPR) test detects antibodies to lipid released by damaged host cells and to lipid in the treponeme's cell wall. Antigen (cardiolipin, cholesterol and lecithin) is mixed with carbon particles that clump in the presence of serum containing antibodies. Antibodies appear 7–10 days after the chancre in 80% of patients and are found in 99% of patients with secondary disease. They decline thereafter or following treatment. A third of late syphilis cases have negative tests. False positive reactions can occur transiently with febrile illnesses or persistently with autoimmune diseases.

Specific serological tests use antigen from *Treponema pallidum* grown in rabbit testes. The haemagglutination assay (TPHA) is used for screening and detects antibody by agglutination of sheep red cells coated with treponemal antigen. Serum has first to be absorbed to remove non-specific reactants. Antibodies are detected in 65% of patients from the fourth week and in almost 100% with secondary disease and patients remain positive throughout the late disease. In disease of more than 2 years' duration, CSF should be obtained since a negative TPHA on CSF excludes neurosyphilis. The fluorescent treponemal antibody absorption test uses organisms fixed on a slide. Antibodies in absorbed serum which bind to them are detected by a fluorescein-labelled anti-human antibody. It is a confirmatory reference test of high sensitivity and specificity which becomes positive 3 weeks after infection and persists after treatment. Yaws and pinta cannot be distinguished from syphilis by serological tests.

17 *Borrelia* Species

Borrelia species are small spiral organisms of the family Treponemataceae. They are transmitted by arthropod vectors and cause relapsing fever (e.g. *B. recurrentis* and *B. duttoni*) and Lyme disease (*B. burgdorferi*). Relapsing fever, a worldwide problem, has been known for years, but the cause of Lyme disease was only discovered in 1975 after a cluster of cases in Lyme, Connecticut.

TRANSMISSION *B. recurrentis* is transmitted by the human body louse (*Pediculus corporis*). The organism does not infect the louse cells but lives within the haemolymph. It is not transmitted by biting but by crushing the louse on the skin. Borrelia are reputed to be able to enter intact skin and mucous membranes.

B. duttoni (and 14 other species) are transmitted from a reservoir in small rodents by argasid soft ticks (e.g. *Ornithodoros moubata*). The organism invades tick cells, may be passed to subsequent generations transovarially and is excreted in the saliva and faeces.

B. burgdorferi is transmitted by hard ticks (e.g. *Ixodes dammini*) and the definitive reservoir seems to be rodents and deer. Larval ticks tend to feed on mice and adults on deer. Both will feed on man given the opportunity.

CLINICAL FEATURES Relapsing fever, is as the name suggests, an acute pyrexial illness characterised by spontaneous remission after about 5 days with a sudden relapse after a further week. Lyme disease is an acute febrile illness with characteristic rash followed after an interval by arthritis, central nervous system (CNS) and skin disease. In many ways, this is like the primary and late manifestations of syphilis.

THERAPY AND PROPHYLAXIS Prevention is by taking steps not to be bitten by ticks or lice. Treatment has been successful with penicillin, erythromycin, tetracycline or chloramphenicol.

LABORATORY DIAGNOSIS In acute relapsing fever, *Borrelia* species can be seen in the peripheral blood. Thick films or dark-ground microscopy should be used. *Xenodiagnosis* (inoculation into mice or pathogen free vectors) has also been used. Lyme disease is usually diagnosed serologically but this has many limitations. These organisms can be grown *in vitro* in enriched media, but this is not generally available.

CLINICAL FEATURES

Relapsing Fever

The patient from an endemic area develops acute fever with rigors, headache, myalgia, arthralgia, and perhaps a transient rash. Patients may have suffused

conjunctivae and even frank bleeding, but the CSF is clear. Infestation with body lice is important though unlikely unless 'living rough' for a period, but ticks can be picked up during short safaris. The patient may remember having a tick bite. The differential diagnosis is from tick-borne rickettsial disease (e.g. R. *conori*). Classically, the fever remits after 5 days or so and then returns after a week, usually without the rash. The cycle may repeat itself three times but on each occasion the fever is less. The illness of louse-borne relapsing fever is more severe with a higher mortality, but rash is more common with tick-borne disease. Treatment will precipitate a Jarisch–Herxheimer reaction (as in syphilis; see previous chapter).

Lyme Disease

The first manifestation is erythema chronicum migrans, starting as a red papule at the site of the tick bite, spreading with a livid red edge while the centre becomes pale or blue. Rings are seen, and the patient may then get satellite lesions at other sites. It may be mistaken for ringworm but the rash is not itchy, the tick bite is usually remembered and fungal scrapings are negative. The rash lasts for between a day and a year. General systemic symptoms of fever, fatigue and headache are common. Other rashes are common as late manifestations of the disease.

Eighty per cent of patients develop arthritis. A migratory polyarthropathy is common early on, similar to the reactive arthritis common to many infections. Later, severe synovitis may ensue. The patient may also develop chronic CNS symptoms, particularly a polyneuritis or radiculitis with a lymphocytic CNS infiltrate. Brief cardiac involvement (atrio-ventricular block) occurs in 10%. Many other systems may be involved.

THE BACTERIA

Borrelia are 10–30 μm long, 0.5 μm wide and have 3–10 spirals. They can be grown in special media. They can be preserved at −70°C or in ticks. They stain well with Giemsa or Gram stains and may be seen in fresh blood under dark-ground microscopy. Inoculation into a mouse and subsequent examination of the mouse blood under dark-ground may revel the organism. Serodiagnosis is not very helpful.

EPIDEMIOLOGY

The occurrence of these diseases in man is very complex and depends on the level of infection in animal hosts, the prevalence of vectors and the likelihood of vectors biting man.

Ticks

Soft ticks (e.g. *Ornithodoros*) tend to infest rodents but they are only resident during meals (5–20 minutes) and do not cause a painful bite. They may go unnoticed on people who pick them up walking through low scrub. They tend to feed at night. There has been an outbreak of tick-borne relapsing fever in campers in America, coinciding with local death of many rodents (an 'epizootic').

Hard ticks (e.g. *Ixodes*) are distinguished by the triangular scutum or shield visible on the dorsal surface of the anterior abdo-thorax and stay on the host for a much longer period, particularly the adult which may take several days to take a blood meal. The larvae need to feed more often for a shorter time in order to mature. The average tick burden on a deer is 6000. In most areas where deer graze, a proportion of the ticks have evidence of infection with *B. burgdorferi*.

Ticks are picked up by walking through low bush or tall grass, particularly with woollen socks and trousers on. The method of counting ticks depends on dragging a blanket over a fixed area of bush land.

Lice

Body lice (*Pediculus corporis*) live in the seams of garments of people who do not change their clothes. Infestation and outbreaks of relapsing fever or typhus are common in times of war and famine and in winter when people congregate together. Head lice (*Pediculus capitis*) and pubic lice (*Phthiris pubis*) are common in clean people, residing in the hair, but do not transmit disease.

THERAPY AND PROPHYLAXIS

Penicillins, erythromycin and tetracycline have all been used successfully in the treatment of both relapsing fever and Lyme disease. Jarisch–Herxheimer reactions may follow antibiotic therapy in both conditions. In Lyme disease symptoms occasionally persist after antibiotic therapy.

Prevention of infections due to *Borrelia* can be achieved by avoiding contact with the relevant insect vector. In terms of lice this involves good personal hygiene whereas for the tick-borne infections the strategy is to avoid the natural habitat of the ticks or wear protective clothing.

LABORATORY DIAGNOSIS

In the acute phase of relapsing fever *Borrelia* can be seen in stained blood smears of more than 70% of cases. The sensitivity of this can be enhanced by inoculating mice with blood samples and then examining for *Borrelia* in the mouse blood after 1–10 days. By contrast direct microscopy of blood or tissues is not a sensitive method for the diagnosis of Lyme disease.

Lyme disease is best diagnosed by serological methods—either immunofluorescence assay or enzyme-linked immunosorbent assays (ELISAs). IgM antibodies develop about 3–6 weeks after infection and IgG antibodies subsequently appear. Some cross-reactions do occur and this together with the delayed appearance of antibodies and the low levels in antibiotic treated cases may make laboratory diagnosis of Lyme disease difficult. There are no useful serological tests for relapsing fever.

18 Leptospiras

Leptospiras are slender spiral organisms with many coils and hooked ends. There are two species *Leptospira interrogans* and *L. biflexa*. The former species contains all the pathogenic leptospiras which are distinguished by serological means. The serotypes which commonly cause disease in man are given specific names, i.e. *L. icterohaemor-rhagiae*, *L. canicola* and *L. hebdomadis*.

TRANSMISSION Animals are the natural host of pathogenic leptospiras (*L. icterohaemorrhagiae*, rat; *L. canicola*, dog; *L. hebdomadis*, mouse). The leptospiras are carried in the kidneys of their animal host and transmission depends upon contact with environments contaminated with infected animal urine. The organisms enter the body through breaks in the skin or through mucous membranes.

CLINICAL FEATURES Many infections are asymptomatic or associated with a mild influenza-like illness. Severe disease with jaundice and haemorrhage (Weil's disease) may occur or the clinical picture may be dominated by signs and symptoms of meningitis.

THERAPY AND PROPHYLAXIS The treatment of choice is penicillin. (This may cause an exacerbation of symptoms, the Jarisch–Herxheimer reaction.) Tetracycline or erythromycin may be used in penicillin-allergic subjects. Infection can be prevented by avoiding occupational or recreational exposure. Immunisation of animals has reduced the reservoir of pathogenic leptospiras for man.

LABORATORY DIAGNOSIS Organisms may be seen by dark-ground microscopy in blood or urine and may be isolated from these specimens using special media. However, the diagnosis is usually made by serology.

CLINICAL FEATURES

Leptospiras gain access to man via abrasions in the skin or through the mucous membranes of the mouth and oropharynx. This is followed by a bacteraemic phase during which particular target organs are seeded. Thus in the early part of the infection bacteria can be found in the blood (and often the CSF even in the absence of signs of meningitis) whereas in the later phases of the infection the bacteria are present in the urine. The major clinical manifestations of the disease are associated with damage to the vascular endothelium of small blood vessels but it is also believed that the immunological response to infection contributes to cell and organ damage. Infection with pathogenic leptospira may result in a subclinical infection, a non-specific febrile illness or a severe systemic disease with particular involvement of

the central nervous system, the liver and the kidneys.

The incubation period is 1–2 weeks and this is followed by a mild illness with fever and myalgia which coincides with the bacteraemic phase of the infection. The symptoms of this phase usually remit after about one week and there may be no further signs and symptoms. Alternatively illness relating to specific target organs, namely the meninges, the liver and the kidney may develop. A mild and benign meningitis may occur and the mortality from this is very low. Alternatively a serious systemic infection associated with a haemorrhagic rash, vascular collapse and impairment of liver and renal function may occur. This is often given the name Weil's disease. The mortality of this condition is approximately 10% and renal failure is the usual cause of death.

THE BACTERIA

Leptospiras are slender spiral organisms approximately 0.1 μm across and 6–20 μm long which are hooked at one or both ends. They do not stain well with ordinary bacterial stains and are best seen by dark-ground microscopy. Leptospiras are obligate aerobes which can be cultured in liquid media supplemented with serum or albumen. There are two species *L. interrogans* and *L. biflexa*, the former containing the organisms pathogenic for man and the latter a group of saprophytic bacteria. *L. interrogans* can be subdivided extensively by serological means into serogroups and serotypes. Many of these are given specific names and those that are particularly associated with disease in man are *L. icterohaemorrhagiae*, *L. canicola* and *L. hebdomadis*.

EPIDEMIOLOGY

Leptospiras are worldwide in distribution. They cause zoonoses in a wide variety of animal hosts. Man is infected coincidentally and is a 'dead end' host. Rodents, dogs and cattle are common hosts and a specific serotype tends to be restricted to a single host animal. For example, the brown rat is the natural host of *L. icterohaemorrhagiae*, the dog the natural host of *L. canicola* and the field mouse the natural host of *L. hebdomadis*.

Man is usually infected by occupational or recreational exposure to water, soil or vegetation contaminated with the urine of animals. Thus there is occupational exposure amongst sewer workers, farmers and veterinarians, and recreational exposure in those who bathe or sail in contaminated waters. The incidence of recognised infection in developed countries is low but this is almost certainly a gross underestimate of the number of infections that take place because many will be subclinical.

THERAPY AND PROPHYLAXIS

Leptospirosis responds to antibiotics administered early in the infection. Penicillin is probably the treatment of choice though tetracyclines and erythromycin have also proved successful. A temporary exacerbation of symptoms may follow the use of antibiotics and this is a Jarisch–Herxheimer reaction analogous to that seen in syphilis.

Leptospirosis can be prevented either by avoiding exposure or by controlling the pool of infection in domestic animals. Agricultural and sewer workers should protect their skin when cleaning out buildings, digging ditches, working in sewers etc., and the rat populations in these situations controlled. Domestic animals can be vaccinated using inactivated whole bacteria or outer membrane proteins and this has been practised extensively in the UK in pet dogs.

LABORATORY DIAGNOSIS

Leptospiras may be identified by dark-ground microscopy in the patient's blood in the first week of the illness or in the sediment of a centrifuged urine sample during the second week. Organisms can be cultured using special media from blood, urine or CSF.

Often, however, the disease is sufficiently non-specific in the early stages to necessitate the use of serology for diagnosis. A variety of tests are available in reference laboratories and diagnosis is based on rising antibody titres or the detection of specific IgM.

19 *Bordetella* Species

Bordetella species are Gram-negative bacilli that are pathogens of ciliated respiratory epithelium in homeothermic animals. There are four species: only two, *B. pertussis* and *B. parapertussis*, are human pathogens.

TRANSMISSION   Infection is spread by droplet nuclei and patients are particularly infectious during the early part of the clinical illness. The incubation period is about one week.

CLINICAL FEATURES   The characteristic illness caused by *B. pertussis* is whooping cough. There is a prodromal catarrhal phase followed by a paroxysmal phase characterised by a repetitive cough, respiratory embarrassment and inspiratory stridor (whoop) and a recovery phase.

COMPLICATIONS   Serious respiratory and neurological complications can occur.

THERAPY AND PROPHYLAXIS   Antibiotics are of no value unless given early in the course of infection. Erythromycin is the antibiotic of choice. Prevention is by vaccination which gives 80% protection. Neurological side effects of infection are far more common than neurological side effects of vaccination and family history of epilepsy is not a contraindication to vaccination.

LABORATORY DIAGNOSIS   Early diagnosis is made by immunofluorescent stains or culture of nasopharyngeal secretions and a diagnosis of infection late in the illness by serology.

CLINICAL FEATURES

Whooping cough is principally a childhood illness. After an incubation period of about a week the primary stage is a catarrhal illness. This is characterised by non-specific signs and symptoms of malaise, and pyrexia, with upper respiratory signs of rhinorrhoea, sneezing and conjunctivitis. Clinically it resembles a viral infection of the upper respiratory tract and during this stage the illness is highly communicable. The catarrhal stage lasts about a week and evolves into the pertussive stage during which the characteristic 'whoop' develops. Frequent repetitive coughs are terminated by an inspiratory gasp ('whoop'). Respiratory embarrassment is obvious and the patient may be cyanotic. Ventilatory support may be required. This severe stage may last up to a month and is followed by a recovery phase where the frequency and severity of paroxyms diminishes.

During the illness there is characteristically an exceptionally high leucocytosis, 200 000 cells/ml, of which about 90% are lymphocytes.

COMPLICATIONS

The illness can be complicated by suppurative infections of the respiratory tract. In the upper respiratory tract otitis media can develop. Because of the retention of viscid secretions, atelectasis may occur and be followed by pneumonia. Aspiration pneumonia can follow a vomiting episode induced by a paroxysm. Bronchiectasis can be a long-term consequence of infection. In addition to the local suppurative complications, the respiratory spasm, by causing venous engorgement, anoxia, raised thoracic and intra-abdominal pressure, can be followed by epistaxis, facial petechiae, cerebral haemorrhage, convulsions, penumothorax hernia and rectal prolapse.

THE BACTERIA

Bordetella are aerobic Gram-negative coccobacilli measuring about 0.4–1.0 μm and microscopically cannot be distinguished from *Haemophilus*. There are four species but only two are of medical importance: *B. pertussis* and *B. parapertussis*.

The organisms are isolated from clinical specimens on Bordet–Gengou medium, or charcoal blood agar. Minute colonies of *B. pertussis* are visible after 3–5 days, incubation, but *B. parapertussis* produces visible growth in 1–2 days. There are a number of cellular components of *B. pertussis* that may act as virulence factors: (1) a cytoplasmic protein that has dermonecrotic properties by inducing arteriolar constriction; (2) a fragment of the peptidoglycan which causes ciliostasis and subsequent destruction of the ciliated cells; (3) lipopolysaccharide which has the characteristic properties of endotoxin; (4) pertussis toxin only produced by *B. pertussis* which produces a wide range of physiological effects on cellular metabolism; (5) an adenyl cyclase which can enter mammalian cells and catalyses the production of increased levels of cAMP; (6) haemolysin (HA); (7) filamentous haemagglutinin; and (8) three surface agglutinins (named 1, 2 and 3). It is not clear how any of these substances relate to the respiratory illness that *Bordetella* produces. However, the surface agglutinins seem to play a role in immunity and all are present in the vaccine.

EPIDEMIOLOGY

Bordetella species are obligatory respiratory pathogens colonising ciliated respiratory epithelia and are found in several warm-blooded animal species.

B. pertussis is a worldwide pathogen, is highly communicable and is transmitted by droplet infection from an index case. The illness has an incubation period of about one week. All age groups are susceptible to infection, but over 80% of all cases occur in those less than 10 years. Adults can develop atypical, or sometimes typical, attacks of whooping cough. The disease is spread easily in overcrowded conditions. With improvements in socio-economic conditions the number of annual notifications in the UK has fallen from 60 000 in 1940 to 5000 in 1991. Most illness is reported in late winter to early spring, and there are peaks of prevalence every 3–4 years. *B. parapertussis* causes a milder illness.

THERAPY AND PROPHYLAXIS

Antibiotics have little role to play in treating cases of whooping cough because, often, by the time the clinical diagnosis is made antibiotics have no effect upon the course of the disease. Erythromycin, if given early, may ameliorate the severity of the illness and it also eradicates the organism from the nasopharynx, thus eliminating a potential source of infection. Erythromycin can be used for prophylaxis of unvaccinated close contacts of a case.

Prevention of infection is by isolation of an index case from susceptible individuals. Illness can also be prevented by vaccination. The current vaccine is a heat-killed whole cell preparation of *B. pertussis*, which gives over 80% protection. The vaccine is given as part of the primary vaccination course (with diphtheria and tetanus) at 2, 3 and 4 months. Side effects occur at a rate of 1:5000 to 1:50 000 and neurological complications at a rate of 1:100 000 which is six times less than the neurological complications following natural infection. A family history of epilepsy or a stable neurological defect are not contraindications to vaccination.

LABORATORY DIAGNOSIS

The difficulty of making a microbiological diagnosis of whooping cough relates to the delay in suspecting clinically that the patient has the illness. By then, attempts to culture the organism from specimens are unrewarding. There are three general methods for establishing a microbiological diagnoses: microscopy, culture and serology. Early infections are best diagnosed by fluorescence microscopy and culture of a perinasal swab, and late infections by serology.

Immunofluorescent staining of nasopharyngeal secretions can provide a rapid answer, although both false-positive and false-negative results do occur. Isolation of *Bordetella* species can be achieved in 80% of cases if the specimens are taken in the catarrhal stage of the illness but isolation rates fall dramatically in the pertussive stage. Pernasal specimens are the method of choice although the organism can also be isolated from 'cough plates', where the plate is held directly in front of the patient during coughing. Freshly prepared Bordet–Gengou or charcoal blood agar should be used. The plates can be made more selective by the addition of antibiotics (penicillin, or cephalexin). Antibodies can be detected by the third or fourth week of the illness and paired sera are necessary to demonstrate a fourfold rise in the titre. Several serological tests are available including an ELISA.

20 *Haemophilus* species

Haemophilus species are small Gram-negative bacilli which can grow under aerobic or anaerobic conditions but only in the presence of factors, X and/or V, derived from blood. *Haemophilus influenzae* is part of the normal respiratory flora but can cause meningitis, epiglottitis, pneumonitis and otitis. *H. ducreyi* is associated with a sexually transmitted disease, chancroid. *H. aegyptius* causes an acute purulent conjunctivitis.

TRANSMISSION   Most *Haemophilus* species are transmitted by infected droplets or contact with respiratory secretions. *H. ducreyi* is spread by sexual contact. The incubation period is usually between 1 and 5 days.

CLINICAL FEATURES   *H. influenzae* serotype b causes a purulent meningitis, epiglottitis or septicaemia in children. Non-encapsulated strains are associated with otitis media, sinusitis and bronchitis. Pneumonia occurs but can be difficult to diagnose. Chancroid is characterised by genital ulcers and inguinal lymphadenopathy. Conjunctivitis is the characteristic illness due to *H. aegyptius*.

COMPLICATIONS   Meningitis may develop following otitis media or pharyngeal infection. Up to 11% of patients have permanent neurological sequelae. Complete airway obstruction can follow examination of the pharynx in epiglottitis.

THERAPY AND PROPHYLAXIS   Resistance to ampicillin is increasingly common. Chloramphenicol or cefotaxime are preferred treatments for severe infections. Rifampicin prophylaxis is used for household contacts under 4 years old in addition to immunisation. In the UK, all children from 2 months of age are given 3 monthly doses of vaccine against *H. influenzae* serotype b.

LABORATORY DIAGNOSIS   *Haemophilus* species can be cultured on media containing X and/or V factors. Blood culture is the most important investigation in all serious infections and urgent direct examination and culture of CSF is obligatory in meningitis.

CLINICAL FEATURES

Strains of *H. influenzae* cause meningitis, epiglottitis, arthritis and pneumonia. *H. influenzae* serotype b is a common cause of meningitis between 1 month and 2 years of age and is of similar frequency to *Neisseria meningitidis* in unimmunised children aged 2 to 6 years. Symptoms and signs are typical of any acute pyogenic meningitis. Colonisation of the nasopharynx precedes invasion of the blood and meninges. Epiglottitis occurs in children between the ages of 2 and 4 years and is often accompanied by septicaemia. The onset is acute with laryngeal oedema that results in airway obstruction. Cellulitis is characterised by the rapid development of a

violaceous swelling in the cheek of a child under the age of 2 years. Septic arthritis affects large joints, e.g. the hip or knee. Pneumonia comprises 3% of adult cases and accompanies systemic infection in children.

With *Streptococcus pneumoniae*, non-capsulated strains of *H. influenzae* are the commonest isolates from cases of chronic bronchitis and bronchiectasis but probably play a secondary role. They are frequently implicated in otitis media and sinusitis, often following pharyngitis or a primary viral infection, and they are isolated in up to 10% of cases of non-gonococcal urethritis. Meningitis caused by non-capsulated strains or serotypes other than b is occasionally found in immunosuppressed patients or secondary to trauma.

H. parainfluenzae has occasionally been isolated from blood or in endocarditis but its presence in patients with pharyngitis usually represents colonisation rather than infection. It is associated with acute exacerbations of chronic bronchitis. *H. ducreyi* causes chancroid, a sexually transmitted disease endemic in South America, Africa and Asia, in which painful ulcers develop on the genitalia and around the anus, together with suppurative inguinal buboes. *H. aegyptius* (Koch–Weeks bacillus) causes an acute purulent conjunctivitis in hot climates.

COMPLICATIONS

Meningeal signs and symptoms may develop following an upper respiratory tract infection or purulent otitis media due to *H. influenzae*. Between 8 and 11% of patients have residual neurological deficits after recovery from meningitis and mortality is 4–5%. Pharyngeal examination of patients with epiglottitis can cause complete airway obstruction and should only be attempted with facilities for immediate intubation and support. Bacteraemia can occur in children without evidence of local disease but is often associated with infection of the meninges, epiglottis, skin, joints, bones, eyes, parotid glands or genitourinary tract.

THE BACTERIA

Haemophilus species are small Gram-negative rods which form filaments or fusiform swellings on prolonged culture. There are six types of capsular polysaccharide which form the basis of serotyping of strains as a to f and non-typable. The capsule confers resistance to phagocytosis and the effects of complement. The invasive type b contains pentose rather than hexose elements. Proteins in the outer membrane can disrupt ciliated bronchial epithelium leading to stasis of mucus and colonisation. Non-typable strains adhere to buccal epithelium by means of pili but these may need to be lost before invasion can take place.

 H. influenzae grows best at 37°C in air with added carbon dioxide. Chocolate (heated blood) or Levinthal agars are used which contain X and V factors necessary for growth. X factor is a protoporphyrin, needed for the synthesis of porphyrins and haem, and V factor is nicotinamide adenine dinucleotide, needed in oxidation–reduction reactions. Encapsulated strains produce mucoid, glistening colonies compared with the small clear colonies of other strains. In blood agar, only the X factor is readily available and clear pinpoint colonies are produced. Their size increases near colonies of staphylococci due to a local production of excess V factor (satellitism). Growth on a nutrient agar near discs containing X, V or X-V factors distinguishes *H. influenzae* from *H. parainfluenzae* which requires only V factor. Serotyping is easily performed by observing agglutination of a suspension of organisms in a drop of saline upon addition of group specific antisera.

 H. ducreyi and *H. aegyptius* are difficult to cultivate and require special media. The former is recognised by finding intracellular and extracellular organisms in Gram-stained smears of ulcer exudate.

EPIDEMIOLOGY

Infection with *H. influenzae* follows inhalation of infected droplets or contact with nasal or pharyngeal discharge from active cases or carriers. Symptomatic infections are much less common than colonisation. The organism is found in the nasopharynx of up to 75% of healthy children yet only 5% of strains are encapsulated and only 2.5% are type b. In adults, 35% are carriers but only 0.4% harbour type b. A relative deficiency of anti-capsular antibody in children between the ages of 3 months and 3 years corresponds with the highest incidence of invasive disease. The risk of spread of meningitis due to *H. influenzae* within a household in children under 5 years is 2.3%, 800 times the risk in the general population. Splenectomised patients, those with sickle cell disease and alcoholics are more susceptible to infection.

Conjunctivitis is spread by contact with discharge from an infected eye or upper respiratory tract usually in children under 5 years old. Chancroid occurs following direct sexual contact with discharge or pus from open lesions.

THERAPY AND PROPHYLAXIS

Resistance to ampicillin is becoming increasingly common among strains of *H. influenzae* and has reached 50% in some parts of Spain. Chloramphenicol is the preferred treatment of meningitis and epiglottitis but ampicillin or co-trimoxazole are appropriate for some conditions, e.g. otitis media. A few strains are resistant to chloramphenicol. Cefotaxime is effective against β-lactamase producing and chloramphenicol-resistant strains and is a popular choice for meningitis in children, as monitoring of serum antibiotic levels is not necessary. Cephalosporins, e.g. cefuroxime, are often used in respiratory infections requiring hospitalisation.

Cases of haemophilus meningitis are isolated for 24 hours after starting treatment. Rifampicin is recommended for chemoprophylaxis for household and day-centre contacts of meningitis where children between 2 and 4 years old have been exposed. Conjugate vaccines against *H. influenzae* type b (Hib vaccine), containing capsular antigens linked to proteins, are effective in over 95% of children immunised. In the UK, three doses are given at monthly intervals to all children from 2 months of age. All unimmunised contacts under 4 years old should receive vaccine.

Chancroid responds to co-trimoxazole or amoxycillin/clavulanate. Patients with chancroid must avoid sexual activity until all lesions are healed. Sexual contacts within 2 weeks of onset should be found and treated.

Conjunctivitis is treated with topical ointment or drops containing tetracycline or chloramphenicol.

LABORATORY DIAGNOSIS

Blood culture is the most important investigation for all serious *H. influenzae* infections. Laryngeal secretions are helpful in epiglottitis but should only be collected if facilities for immediate ventilation are available. CSF, joint aspirate, pus and throat swabs can be useful but, in bronchitis and pneumonia, bronchial washings are more likely than sputum samples to diagnose infection rather than colonisation.

Samples should be transported without delay to the laboratory and transport medium must be used for swabs as the organisms are easily killed by drying. Samples from genital ulcers or the conjunctivae should be inoculated directly onto media. In meningitis, direct examination of a Gram-stained smear and culture of a spun deposit of CSF must be performed. *H. influenzae* type b appears as short, often scanty, Gram-negative rods and immediate confirmation may be possible by latex agglutination which uses particles coated with monoclonal anticapsular antibodies. If mixed with type b antiserum and methylene blue on a slide, the appearance of a halo around the organisms after 10 minutes is diagnostic and is caused by a change in the refractive index of the organism (Quellung reaction). Production of β-lactamase can be detected rapidly by hydrolysis of a chromogenic cephalosporin in filter paper but antibiotic susceptibility testing is essential to determine treatment. Chancroid can be diagnosed by the typical microscopic appearance of *H. ducreyi* in Gram-stained smears of ulcers or pus.

21 *Legionella*

Legionella species make up a metabolically fastidious group of Gram-negative organisms that are found widely in warm, moist environments and principally cause respiratory infections. There are over 30 species but the main human pathogen is *L. pneumophila*.

TRANSMISSION   The organisms can be found in environmental or man-made water sources where they are often found in symbiosis with various species of amoeba. The mode of transmission is most likely by inhalation of contaminated aerosols often disseminated by air-conditioning systems.

CLINICAL FEATURES   The principal illnesses caused by *Legionella* are the respiratory infections of Pontiac fever, a mild illness with upper respiratory tract signs and myalgia, and Legionnaire's disease, a pneumonic illness with a significant mortality.

THERAPY AND PROPHYLAXIS   Control of epidemics relies upon adequate maintenance and treatment of potable water supplies and air conditioning systems, particularly in public locations. Treatment of established infection is with erythromycin or erythromycin and rifampicin.

LABORATORY DIAGNOSIS   Legionellas can be isolated from bronchial washings, sputum and blood. A diagnosis can also be established serologically by detecting an antibody response or detection of legionella antigens in the urine.

CLINICAL FEATURES

Legionella infections can occur in all ages. However, clinically significant disease is highest in males over the age of 55. Infection may occur in otherwise healthy individuals, but is more likely in association with certain risk factors such as smoking, chronic obstructive airways disease, diabetes, emphysema and immunosuppression, e.g. chemotherapy and transplantation.

Although asymptomatic infections can occur, the main clinical presentation is that of Pontiac fever and Legionnaire's disease.

Pontiac fever is a mild, febrile illness with a low mortality. The patients are pyrexial and complain of myalgia, headache, sore throat and cough, developing over a period of 6–24 hours. Respiratory signs and symptoms are minimal. Meningeal signs such as neck stiffness and photophobia can occur and the patients may be confused. The

illness is short and resolves over a period of a few days.

Legionnaire's disease is an acute onset, principally respiratory illness of variable severity which may have a significant mortality rate of about 10–20%. The patients develop a high temperature with rigors, a non-productive cough and dyspnoea. Pathologically the distal respiratory airspaces are filled with a fibrinous exudate containing numerous polymorphonuclear neutrophils and macrophages. Nodular inflammatory lesions occur commonly. Chest X-ray shows a patchy consolidation which may extend to a multilobar distribution. Abscesses may also be visible. The patient is hypoxic and may have signs of extrapulmonary involvement. Hyponatraemia and a leucocytosis are common. Haematuria may occur and renal failure can supervene. There is evidence of liver involvement with deranged liver function tests. The patient may also be confused and delirious. Abscesses have been detected in liver, spleen and the intestines.

THE BACTERIA

The genus *Legionella* currently comprises more than 30 species although the main human pathogen is *L. pneumophila*. All are fastidious aerobic Gram-negative bacilli, measuring 0.5–1. × 2–4 μm, but on artificial media can produce filamentous forms up to 50 μm long. They can grow over a wide temperature range and have been isolated from water sources at 63°C. The organisms are relatively resistant to acid and chlorine. They stain poorly with Gram's stain but can be detected by using silver stains. Legionellas are relatively unreactive biochemically and their classification and species identification relies upon chemotaxonomy of their fatty acid and ubiquinone content, serology and studies of DNA relatedness.

Legionellas have a complex antigenic structure, and *L. pneuomophila* can be subdivided into 14 serogroups. Each serogroup can be further subdivided into serotypes, based upon the variability of the sugar side chain on the cell wall lipopolysaccharide. The other species of *Legionella* can also be divided into serogroups. Most *Legionella* species produce β-lactamases.

Legionellas are intracellular pathogens and produce a number of extracellular products, such as lipases, proteases and DNAases, some of which may be involved in lung damage. A cytotoxin has been demonstrated which inhibits phagolysosome fusion and superoxide generation in phagocyte cells, thus ensuring the organism's intracellular localisation to alveolar macrophages.

EPIDEMIOLOGY

Legionellas have an extensive environmental reservoir and are found in freshwater lakes and streams, hot and cold water domestic plumbing systems as well as those of hospitals and hotels and air conditioning systems. Certain components of potable water supplies, e.g. black rubber washers, predispose to colonisation by legionellas. The organisms can live intracellularly in various protozoa that are found both in the environment as well as in plumbing and air conditioning systems.

Cases of Legionnaire's disease may be sporadic, but epidemics do occur, often associated with air conditioning systems, affecting the general public, or hospital or hotel populations. The main route of infections is most likely by inhalation of aerosols of contaminated water, although ingestion of contaminated potable water is also a potential route of infection.

The species most frequently giving rise to infection, both Legionnaire's disease and Pontiac fever, is *L. pneumophilia*, serogroup 1. Some other species of *Legionella* have also been isolated from human infections.

THERAPY AND PROPHYLAXIS

Treatment is with erythromycin in the first instance but it may be several days before the patient responds. Naturally erythromycin-resistant isolates have not been detected although resistance can be induced in laboratory experiments. If the patient is severely ill, or if the patient does not respond to erythromycin alone, then a combination of erythromycin and rifampicin should be used. Doxycycline (or ciprofloxacin) is an alternative if the patient cannot tolerate erythromycin. *In vitro* susceptibility is not predictive of clinical response and requests for routine antibiotic sensitivities are not needed.

LABORATORY DIAGNOSIS

Legionellas are usually recovered from broncho-alveolar lavage, sputum or blood specimens and lung biopsies. They can be detected directly in specimens using immunofluorescent stains or DNA probes. These rapid direct detection methods are likely to become more important in diagnostic protocols. Legionella antigen can be detected in urine. The results are obtained rapidly and they have a sensitivity similar to that of culture.

The bacteria may be cultured in BCYE (buffered charcoal yeast extract) media with a sensitivity of above 80%. Detection using culture methods may take from 3–4 days to as long as 1–2 weeks with blood cultures. Saline inhibits the growth of *Legionella* and this should be remembered when performing broncho-alveolar lavage.

Legionella infection can be detected serologically using a rapid micro-agglutination test (RMAT) in which a fourfold rise in antibody titre in sequential specimens, or a titre of 256 in a single specimen from a patient with a suggestive clinical history is indicative of infection. In some cases the serological response may be delayed for some weeks. The antibody response may be prolonged for many years and up to 12% of a healthy population can have a titre of 256.

22 Neisseria (and Moraxella)

The organisms are Gram-negative cocci or coccobacilli. They are non-motile, aerobic, oxidase and usually catalase positive. They are strictly parasitic with man as the natural host. They often require enriched, serum-supplemented media for growth in the laboratory. Two species of *Neisseria* are of importance in medicine, *N. meningitidis* which causes septicaemia and meningitis, and *N. gonorrhoeae* which causes gonorrhoea.

TRANSMISSION   Transmission of *N. meningitidis* is via the respiratory route between close contacts, e.g. family members and school children. *N. gonorrhoeae* is spread exclusively via genital contact, including infection of the eyes of newborn babies. The hands may act as a secondary means of transmission.

CLINICAL FEATURES   Meningococcal infections are characterised by a purpuric rash, with profound toxaemia. Signs of meningitis may or may not be present. Gonorrhoea causes a purulent urethral or cervical discharge. Involvement of the upper female genital tract leads to pelvic inflammatory disease (salpingitis).

THERAPY AND PROPHYLAXIS   *N. meningitidis* is sensitive to penicillin, chloramphenicol and third generation cephalosporins. *N. gonorrhoeae* is also usually sensitive to penicillin, but β-lactamase producing strains are widely reported. Second generation cephalosporins, fluorinated quinolones and spectinomycin are widely used.

A vaccine is available against *N. meningitidis* types A and C, but not for type B. No vaccine exists for *N. gonorrhoeae*.

LABORATORY DIAGNOSIS   Gram stain of genital discharge is useful for the presumptive diagnosis of gonorrhoea. Aerobic conditions supplemented with CO_2 are required for culture of *Neisseria* species of serum-enriched media with added antimicrobials to prevent overgrowth by commensal organisms. Specimens for *N. gonorrhoeae* are best inoculated onto media in the clinic.

CLINICAL FEATURES

Septicaemia and Meningitis

Not all cases of meningococcal septicaemia proceed to meningitis. Onset may be insidious or sudden with rapid deterioration. The incubation period is uncertain, but probably of the order of 2 to 3 days. The patient is febrile and frequently drowsy and

photophobic. Neck stiffness and other signs of cerebral irritation including severe headache and vomiting may be present. The rash, which may be scanty or florid, is purpuric and affects the skin and mucous membranes, including the tarsal conjunctiva. A haemorrhagic rash incidates a poor prognosis.

Gonorrhoea

In men, gonorrhoea presents as a profuse urethral discharge. Symptoms are usually less obvious in cervical infection of women, but mucopurulent cervicitis and increased vaginal discharge occur. Gram stain reveals the presence of intracellular organisms, distinguishing the condition from non-gonococcal or non-specific genital infection. Asymptomatic carriage is a feature of gonococcal infection in both men and women. Gonococcal ophthalmia neonatorum follows birth through an infected cervix, appearing within 3 to 4 days of birth.

COMPLICATIONS

Pelvic inflammatory disease results when infection spreads to involve the uterus (endometitis), the fallopian tubes (salpingitis), and the peritoneum. The latter may result in perihepatitis (Curtis–Fitz-Hugh syndrome, inflammation of the hepatic peritoneum with thin intraperitoneal adhesions). In men, epididymitis and acute prostatitis may occur. Bacteraemic spread (commoner in women) leads to disseminated gonococcal infection (DGI), causing septic arthritis, a skin rash that can be macular, vesicular or purpuric, and rarely myopericarditis and meningitis. DGI may occur in the infected newborn, many of whom are born prematurely.

THE BACTERIA

Neisseria species are Gram-negative cocci, usually associated in pairs with the opposed sides flattened. They are aerobic, and fastidious, requiring media enriched with serum for replication *in vitro*. They are non-motile, and catalase and oxidase positive. *N. meningitidis* is subdivided into a number of serogroups, of which A, B, C and W135 are the commonest. Type B is the predominant strain in the UK. Classification of *N. gonorrhoeae* along the same lines is less satisfactory.

EPIDEMIOLOGY

Some 10% of the population carry *N. meningitidis* in the nasopharynx, and the majority of infections are subclinical. The rate of carriage rises before an outbreak to 20–30%. Outbreaks are particularly likely in enclosed communities. *N. gonorrhoeae* is spread exclusively via sexual contact, both homosexual and heterosexual. Asymptomatic infection is common, up to 50% in some studies, but generally of the order of 10–25%. The hands may act as a secondary means of transmission. Infants born to infected mothers are likely to develop ophthalmia neonatorum.

THERAPY AND PROPHYLAXIS

N. meningitidis is sensitive to penicillin, and this remains the drug of choice, with chloramphenicol, cefotaxime or ceftazidime as alternatives. Sulphonamides were widely used, but increasing resistance means they can no longer be recommended. *N. gonorrhoeae* is also usually sensitive to penicillin, but β-lactamase producing strains are widely reported, and may predominate in certain parts of the world. First line therapy is with ampicillin or amoxycillin if β-lactamase strains are unusual (less than 5%). A second generation cephalosporin (cefuroxime or ceftriaxone), or a fluorinated quinolone (ciprofloxacin or ofloxacin), or the aminocyclitol spectinomycin are used in other circumstances. It is important to treat uncomplicated lower genital tract gonorrhoea with a single dose of antibiotic if possible.

Traditionally, 1% silver nitrate drops (Credé's prophylactic drops) were instilled in the eyes of newborn infants. This practice is unnecessary in areas of low prevalence, and when indicated, erythromycin may provide optimal prophylaxis against both *N. gonorrhoeae* and *Chlamydia trachomatis*.

A vaccine is available against *N. meningitidis* types A and C, but not for type B (the predominant serotype in the UK). No vaccine exists for *N. gonorrhoeae*.

LABORATORY DIAGNOSIS

Gram stain of genital discharge is useful for presumptive diagnosis of gonorrhoea, but results should always be confirmed by culture. Neisseriae require aerobic conditions supplemented with 10% CO_2 for culture, and a serum enriched medium with added antimicrobials to prevent overgrowth by commensal organisms is necessary. Colonies (at 48 h) are 1 mm and grey. A positive oxidase test gives a presumptive identification as *Neisseria* species. Sugar fermentation tests (glucose positive for *N. gonorrhoeae* and glucose and maltose for *N. meningitidis*) give species identification. Rapid confirmation of the identify of *N. gonorrhoeae* can be performed with a co-agglutination slide test.

Moraxella (Some Strains Previously Called *Branhamella*)

These are Gram-negative cocci or coccobacilli. They are aerobic and usually oxidase and catalase positive. *M. (Branhamella) catarrhalis* is primarily a commensal of the upper respiratory tract, but increasingly shown to have a pathogenic role in exacerbations of chronic bronchitis. *M. lacunata* (Morax–Axenfeld bacillus) is a cause of angular conjunctivitis.

CLINICAL FEATURES

M. catarrhalis is one of a number of organisms found commensally in the upper respiratory tract, and also associated with lower respiratory tract infection. Its precise role is difficult to determine, but the organism often produces β-lactamase, and therefore its presence may compromise the therapy of both upper and lower respiratory tract infections with β-lactamase sensitive antibiotics. *M. lacunata* is found either free or in the polymorphs or epithelial cells in cases of acute angular conjunctivitis.

THERAPY

β-lactamase stable β-lactam antibiotics are effective against *M. cararrhalis*, for example second generation cephalosporins such as cefuroxime. Chloramphenicol is effective against ocular infections due to *M. lacunata*.

LABORATORY DIAGNOSIS

M. catarrhalis is less fastidious than other members of this group, and will grow on nutrient agar. The tributyrin test gives presumptive identification. *M. lacunata* grows well on Loeffler's serum medium, producing characteristic pitting and liquefaction.

23 *Brucella* Species

Brucellas are small, Gram-negative coccobacilli which grow aerobically but relatively poorly on ordinary media (the presence of CO_2 enhances growth). All six species of the genus are pathogenic in animals, and at least three species, *Brucella melitensis, B. abortus* and *B. suis*, are pathogenic in man. They may present acutely with fever, lassitude and general debility sometimes associated with bone pain, or may pursue a more long-term, chronic remitting course.

TRANSMISSION   Brucellas are usually transmitted to man by infected genital secretions or fetal material from infected animals. Person-to-person transmission is very rare although occasionally laboratory-acquired infections do occur.

SYMPTOMS AND SIGNS   The incubation period of acute brucellosis is usually some weeks or even months. The infection is characterised by debility, malaise, chills, fever, headache and tiredness, and may last for months. Relapse is frequent and chronicity is common.

COMPLICATIONS   Acute infection may give way to subacute or chronic infection characterised by prolonged malaise, lethargy, debility and ill-health; occasionally osteomyelitis supervenes.

THERAPY AND PROPHYLAXIS   The mainstay of treatment is with antibiotics which penetrate to the intracellular location of the bacteria. Tetracyclines are commonly used sometimes with streptomycin. Co-trimoxazole and rifampicin plus trimethoprim have also been used.

LABORATORY DIAGNOSIS   *Brucella* species grow slowly, usually after a lag phase of several days and none of the pathogenic species produce heavy growth *in vitro*. The addition of CO_2 enhances the growth and may be essential for the isolation of *B. abortus*. Diagnosis of brucellosis is by isolation from blood cultures or other infected tissues, or by a diagnostic rise in specific antibody titres.

CLINICAL FEATURES

Brucellosis is mainly an occupational disease of agricultural workers and those working in slaughter yards, and is acquired by contact with cattle or by the drinking of unpasteurised milk from sheep or goats. Occasional infections in laboratory workers also occur. Infection may be subclinical or lead to a disease which may be acute, subacute or chronic. Acute brucellosis occurs after an incubation period of up to 3 weeks and is rather more common with *B. melitensis* than with *B. abortus*. Main features are high fever, headache, sweating, tiredness and pains in joints; there may be

enlargement of the spleen and hepatomegaly. In untreated patients the disease may become chronic, particularly with *B. abortus*, and after a month is mainly characterised by repeated attacks and extreme tiredness; the joints may be swollen and painful.

COMPLICATIONS

Chronic brucellosis may last for months or years and is characterised by repeated episodes of flu-like illnesses, lassitude, tiredness and depression. Very rarely brucellosis may be complicated by endocarditis, meningitis or chronic arthritis. Granulomas may appear in the reticulo-endothelial system in chronic brucellosis.

THE BACTERIA

Brucella species are small, non-motile, non-sporing Gram-negative coccobacilli. Poor growth on ordinary media occurs aerobically. They are non-fermenters of carbohydrates and do not produce indole. *B. abortus, B. suis* and *B. melitensis* cause infections in man; the other species, *B. canis, B. ovis* and *B. neotomae*, have only very occasionally been transmitted to man. Growth is intracellular and the organisms readily resist phagocytic ingestion. The organisms grow slowly, usually after a lag phase of some days, and with none of the species is growth profuse.

Serum or other animal protein enhances growth in the laboratory and is often added to culture media such as liver infusion broth, trypticase soy agar or glycerol agar, all with serum addition. Growth occurs between 20 and 40°C and with an optimum at 37°C. The addition of CO_2 enhances growth and is needed for the isolation of *B. abortus*.

EPIDEMIOLOGY

Both domesticated and wild mammals are susceptible to brucella infection. *B. abortus* infects cattle in many countries but is now very rare in the UK following a policy of testing and slaughter of infected animals. *B. melitensis* infects goats and is excreted in their milk. *B. suis* is a zoonosis in pigs mainly in the USA. Those working with infected animals such as veterinary surgeons, agricultural workers and slaughter-house staff are prone to infection from aborted materials, and so are those who handle meat or drink unpasteurised milk from infected goats or sheep, or eat products made from untreated milk. It is probable that under-diagnosis is common, particularly in agricultural districts.

Those working in diagnostic laboratories may become infected as a result of contact with infected materials.

THERAPY AND PROPHYLAXIS

The best prevention is by control of brucella infections in domesticated livestock. Tetracycline, usually administered orally for up to 3 months, together with daily streptomycin for the first weeks, is the best treatment for acute brucellosis. Sometimes repeated courses are needed. Relapse is not infrequent because of the intracellular location of the bacteria, which also makes access by antibiotics difficult. Co-trimoxazole or combinations of rifampicin and trimethoprim have also been used successfully. Chronic brucellosis may be exceptionally difficult to treat satisfactorily with antibiotics.

LABORATORY DIAGNOSIS

The blood count in acute brucellosis may be either normal or low with a characteristic neutropenia and lymphocytosis. Blood cultures are the most important individual investigation, being positive in up to 20% of cases, usually during the acute phase of the illness. *B. abortus* requires a high CO_2 concentration. If brucellosis is clinically suspected, diphasic blood culture medium (Castaneda) may be used as an improved method of producing positive results. Liver biopsy may be helpful in isolation of the organism or for histological diagnosis. Serological diagnosis is made by the use of the standard agglutination test, by a complement fixation test or by the anti-human globulin test (Coombs test). At least a fourfold rise in antibody titre, usually collected at least 10 days apart between first and subsequent specimens, is needed to establish the diagnosis. In single samples the presence of IgM antibodies is diagnostic of acute infection.

24 *Francisella, Yersinia* and *Pasteurella* Species

Francisella tularensis causes tularaemia and *Yersinia pestis* causes plague. *Yersinia enterocolitica* and *Yersinia pseudotuberculosis* are associated with diarrhoea and mesenteric adenitis. Infection with *Pasteurella* species causes wound infection following animal bites.

Tularaemia

CLINICAL FEATURES

Tularaemia is associated with the sudden onset of fever, chills and malaise and one of five syndromes, depending on the site of entry. In the ulceroglandular, oculoglandular, oropharyngeal and glandular forms, painful local lymphadenopathy accompanies a skin ulcer, purulent conjunctivitis, exudative tonsillitis or no local lesion respectively. In the typhoidal form, fever, weight loss and shock occur without lymphadenopathy. Pneumonia is found with any form and is suggested by a dry cough, few physical signs and patchy radiological infiltrates.

COMPLICATIONS

Lobar pneumonia, pleural effusion or lung abscess can develop in tularaemia and rash is observed in up to one-fifth of patients. Liver function may be abnormal. Mortality is 1–3%.

THE BACTERIA

Francisella tularensis are very small Gram-negative bacilli which require oxygen and cysteine for growth. They are intracellular parasites and are taken up by macrophages. Virulent organisms have a thick capsule. On glucose–cysteine blood agar, maximum growth is obtained after 2–4 days at 37°C with a green discoloration of the medium.

EPIDEMIOLOGY

Tularaemia is spread throughout the northern hemisphere (except the UK) and is found in many animal species but rabbits, hares, ticks and muskrats are the most important reservoirs for human disease. The infection can be acquired by contact with body fluids, bites, inhalation of aerosols or consumption of contaminated water or meat. Tick-borne cases are most common in summer and rabbit-associated ones in winter. There are numerous reports of laboratory workers having been infected by inhalation of an aerosol of the organism.

THERAPY AND PROPHYLAXIS

For tularaemia, aminoglycosides, particularly streptomycin, are given for 7–14 days. In endemic areas, gloves should be worn if skinning wild animals and tick-infested areas avoided or insect repellent used. A vaccine provides only partial protection. Cell-mediated immunity after infection lasts for life.

LABORATORY DIAGNOSIS

Francisella species can be isolated from the ulcer or an aspirate of the local lymph nodes on cysteine-containing media but handling in the laboratory is hazardous. Blood cultures are not helpful. Immunofluorescence of tissue smears is preferable. Most infections are diagnosed by a fourfold rise in antibody titre, using an enzyme-linked immunosorbent assay. Antibody response is detectable in the second week of illness and peaks at 4 weeks.

Yersinia

CLINICAL FEATURES

In bubonic plague caused by *Yersinia pestis*, the patient develops fever, chills and headache one week after a flea bite, accompanied by an intensely painful swelling of the local lymph nodes (bubo) with surrounding oedema. The groin is most frequently affected. Patients are hypotensive or shocked and may die in 2–4 days. Plague can occur without a bubo (septicaemic plague) when the mortality is 33%. Cough, pleuritic pain and radiological changes suggest pneumonic plague which is rapidly fatal.

Infection by *Y. enterocolitica* causes diarrhoea, abdominal pain and fever for 1–3 weeks associated with ulceration of the ileum and mesenteric lymphadenopathy. In older children this may resemble the symptoms of appendicitis. *Y. pseudotuberculosis* causes a mesenteric adenitis with fever and right iliac fossa pain.

COMPLICATIONS

Pustules or papules may develop in bubonic plague at the site of the flea bites. Purpura caused by vasculitis can result in gangrene of the limbs (Black Death). Meningitis and pharyngitis also occur. *Y. enterocolitica* infection can result in perforation of the ileum and rectal bleeding. Septicaemia occurs in patients with immune suppression, haemolytic anaemia or iron overload, especially thalassaemia. A reactive polyarthritis or erythema nodosum is present in up to one-third of patients for 1–4 months.

THE BACTERIA

Yersinia species are small Gram-negative coccobacilli which do not form capsules but *Y. pestis* can have an envelope. They are aerobic and facultatively anaerobic and grow on simple media, even at 4°C. At 30°C they are motile except *Y. pestis*. Virulence is particularly associated with a plasmid coding for certain antigens (V, W), with the presence of an envelope antigen (F1), purine synthesis and toxin production. The F1 antigen confers resistance to killing by phagocytes.

EPIDEMIOLOGY

Most cases of plague are reported from Africa and Asia, especially India and Vietnam. It is transmitted by the bite of the rat flea or by consumption of infected tissue. The rat is the major reservoir, human cases being accidental infections. If the flea bites an infected animal, the organism produces clotting of the blood in the foregut of the flea. When the flea attempts another meal, organisms are regurgitated into the wound and pass to regional lymph nodes. They are engulfed by mononuclear cells but multiply within the cells, leading to lysis. In pneumonic plague, the disease may be acquired by aerosol inhalation.

Y. enterocolitica is transmitted by contaminated food, particularly pork, milk and water, and will grow in refrigerated meat. It is found in Europe and the USA in a wide variety of domestic animals and rodents. *Y. pseudotuberculosis* has a reservoir in rodents, rabbits, domestic animals and birds and usually infects children, via the gastrointestinal tract, by contact or via food or water.

THERAPY AND PROPHYLAXIS

Plague has a high mortality unless treated early in the disease, usually with streptomycin for 10 days and intravenous fluids. Tetracycline is an alternative. Chloramphenicol is used for meningitis. Patients with pneumonia must be source isolated for 48 hours after the start of treatment. Gloves must be used when handling infected material. Contacts should be treated. A killed vaccine is protective for those entering endemic areas and for laboratory workers. Improvements in sanitation and waste disposal and rat-proofing of houses reduce the risk of spread.

Antibiotic treatment is not necessary in most case of *Y. pseudotuberculosis* and *Y. enterocolitica* infection although aminoglycosides, ampicillin (if sensitive), chloramphenicol or tetracycline can be used. Early and prolonged treatment is needed in cases of septicaemia. Prevention is by adequate cooking of meat, especially pork, avoidance of contact with infected animals and not refrigerating meat for prolonged periods before consumption. Vaccines are not widely available.

LABORATORY DIAGNOSIS

In plague, an aspirate of the bubo should be obtained by injecting and aspirating 1 ml of saline and preparing a smear for Gram and methylene blue stains. Other *Yersinia* species may be cultured from faeces, lymph node biopsy, blood or peritoneal fluid. *Yersinia* species show bipolar staining. On culture, colony sizes increase to 3 mm after incubation for 48 hours. Cultures can be enriched by incubation in buffer at 4°C for 2 weeks but detection in faeces is usually possible by standard techniques.

A serological test, passive haemagglutination, is available for plague and a rising titre of antibodies to one of the O serogroups (I–VI) can be used to diagnose *Y. pseudotuberculosis* infection.

Pasteurella

CLINICAL FEATURES

Infection with *Pasteurella multocida* is suggested by local erythema, swelling and pain within 24 hours of an animal bite. Serous discharge and local lymphadenopathy develop. The organism may be isolated from patients with chronic respiratory disease probably as the result of upper airways colonisation.

COMPLICATIONS

Osteomyelitis, arthritis, tenosynovitis and bacteraemia can occur after animal bites become infected with *P. multocida*.

THE BACTERIA

Pasteurella species are small Gram-negative pleomorphic bacilli. Bipolar staining is common and they are not motile. They are facultative anaerobes and can grow on nutrient media. *P. multocida* forms a capsule that might be associated with virulence and is the basis for serotypes A, B, C and D, the latter being capable of toxin production.

EPIDEMIOLOGY

P. multocida is a common component of the nasopharyngeal flora of domestic and wild animals and most infections follow animal bites, particularly by cats. Respiratory infections usually involve patients with frequent animal contact.

THERAPY AND PROPHYLAXIS

Animal bite wounds should be rapidly cleansed and debrided and pasteurella infection treated with penicillin V, tetracycline or amoxycillin–clavulanic acid. Bacteraemia, osteomyelitis and other deep infections require parenteral antibiotics and surgical drainage.

LABORATORY DIAGNOSIS

Diagnosis of pasteurella infection depends on the isolation of small bipolar Gram-negative bacilli from animal bite wounds, sputum or pus. Strains are identifiable by morphology and biochemical reactions, e.g. production of acid but not gas from glucose and sucrose and positive tests for indole, catalase, oxidase and H_2S.

25 Coliforms, Pseudomonads and Allied Organisms

There are approximately 24 genera of coliforms belonging to the family Enterobacteriaceae. Some (*Escherichia, Salmonella, Shigella* and *Yersinia*) will be covered in other sections of this book because they have an epidemiology and specific clinical syndromes associated with them that are different from the bacteria considered in this section. Of the group of coliforms considered here, *Citrobacter, Enterobacter, Klebsiella, Morganella, Proteus, Providencia* and *Serratia* are the most frequently occurring genera found in clinical samples.

The group of pseudomonads and allied organisms all have a similar type of oxidative metabolism and do not belong to the Enterobacteriaceae. In all, there are about 17 genera, of which *Acinetobacter, Alcaligenes, Agrobacterium, Flavobacterium, Pseudomonas* and *Xanthomonas* are the most frequently occurring in clinical samples.

TRANSMISSION   Coliforms and pseudomonads are part of the normal flora of the gastrointestinal tract and infections may therefore be endogenous. They are also found in a wide variety of environmental sites and are part of the resident hospital environmental flora. Transmission to patients may be direct or indirect from staff, equipment, disinfectants or intravenous fluids.

CLINICAL FEATURES   Septicaemia is the most important clinical illness caused by these organisms. In addition they may cause pneumonia (*Pseudomonas, Klebsiella*); urinary tract infection (*Klebsiella, Proteus*); meningitis (*Klebsiella, Flavobacterium*) or wound infections (many species).

THERAPY AND PROPHYLAXIS   Many of these organisms are resistant to commonly used antibiotics and treatment is guided by sensitivity testing. Anti-pseudomonal β-lactams and aminoglycosides are usually given. Isolation of patients infected with resistant organisms cuts down transmission.

LABORATORY DIAGNOSIS   Septicaemia and focal infections are diagnosed by isolating the organisms from appropriate specimens (blood, urine, sputum etc.).

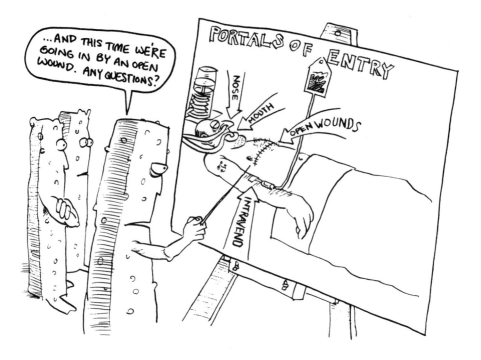

CLINICAL FEATURES

The most important clinical illness caused by these organisms is septicaemia. Compromised patients, especially immunocompromised patients, who become colonised from the hospital environment, or who carry these organisms as part of their resident microflora, may at some stage in their illness develop septicaemia and perhaps septic shock. This is particularly likely to happen if the patient is neutropenic. Entry into the blood is via the gastrointestinal, respiratory or renal tract or the skin, often because of breaches of primary defence barriers associated with instrumentation. Septicaemia with these organisms carries a significant mortality (20–30%). In cases of septicaemia caused by *Ps. aeruginosa* focal skin necrosis, called ecthyma gangrenosum, may occur. The coliforms and pseudomonads also cause pneumonia (*Pseudomonas, Klebsiella*); urinary tract infections (*Klebsiella, Proteus*); meningitis (*Klebsiella, Flavobacterium*) or wound infections and these are often associated with endotracheal intubation, catheterisation, parenteral nutrition and operations.

In addition to nosocomially acquired infections several of these bacteria cause primary infections in other settings. *Klebsiella pneumoniae* is a cause of primary lobar pneumonia, particularly in alcoholic patients. *Proteus* causes about 5–10% of uncomplicated cases of pyelonephritis. *Pseudomonas aeruginosa* causes a wide spectrum of clinical illness. Mucoid strains of *Ps. aeruginosa* are particularly likely to cause pneumonia in patients with cystic fibrosis and these patients also develop pulmonary infections with *Ps. cepacia*. *Ps. aeruginosa* can give rise to a severe destructive endophthalmitis associated with the use of contaminated contact lenses, a destructive otitis externa and a widespread folliculitis associated with the use of jacuzzis.

Pseudomonas pseudomallei causes an illness called melioidosis which may present as a septicaemic illness with widespread skin and deep tissue abscesses following infection acquired in the Far East. Colonisation of the patient can occur many years (15–20 years) prior to the development of clinical illness. *Pseudomonas pseudomallei* may also present in a similar manner to tuberculosis with fever, weight loss, night sweats and cavitating lesions in the upper lobes of the lungs.

THE BACTERIA

The coliforms, pseudomonads and allied organisms are all Gram-negative bacilli. They have a variety of structural and biochemical properties which serve to distinguish the genera and species. For example the coliforms are facultative anaerobes but the pseudomonads are obligate aerobes. Many are motile but klebsiellae are not; the latter have thick polysaccharide capsules. *Pseudomonas aeruginosa* often produces diffusible pigments.

The core region of the cell wall of these bacteria consists of lipopolysaccharide called endotoxin. This plays an important role in the pathogenesis of septic shock since it activates a number of pathophysiological processes some of which lead to cell death and necrosis. Other factors which play a role in pathogenesis are fimbriae, which mediate surface attachment, and extracellular enzymes (lipases, DNAases, proteases, elastase) produced by *Proteus, Pseudomonas* and *Serratia*.

The coliforms and pseudomonads are not fastidious in their nutritional requirements and grow in simple media over a wide temperature range from 0 to 40°C.

EPIDEMIOLOGY

Coliforms and pseudomonads are found as part of the microflora of the gastrointestinal tract. *Acinetobacter* is a skin commensal in about 25% of individuals. All of these bacteria are also found naturally in a wide variety of environmental reservoirs such as water supplies, soil, sewage and vegetation as well as contaminating meats and fish. *Agrobacterium* is found particularly associated with plants and is a recognised plant pathogen. *Pseudomonas pseudomallei* is found in abundance in moist soils in Southeast Asia and northern Australia. With the exception of *Ps. pseudomallei* the remaining bacteria can be found frequently in the hospital environment, where they principally colonise moist areas and may be found contaminating medical equipment or even disinfectant solutions.

On admission to hospital, patients become colonised with the indigenous hospital environmental flora, of which this group of Gram-negative organisms forms a significant proportion. The patients may become colonised directly from the environment; from hands of staff members, who may carry these organisms on their hands and transmit them from patient to patient (e.g. *Klebsiella*); from medical equipment such as ventilators (e.g. *Pseudomonas, Klebsiella, Serratia*); from enteral nutrition (e.g. *Klebsiella, Enterobacter*); from disinfectants (e.g. *Pseudomonas*) or even from contaminated infusion fluids (e.g. *Pseudomonas, Enterobacter*). Many of these bacteria may be found as part of the patient's own microflora and endogenous infections can occur (e.g. *Acinetobacter*).

Pseudomonas pseudomallei is not a nosocomial pathogen and primary infection of individuals occurs by the organism gaining entry to the body through minor skin abrasions after contact with contaminated soil and vegetation in Asia or northern Australia.

THERAPY AND PROPHYLAXIS

The antimicrobial sensitivity pattern of this group of organisms is very variable and treatment should be guided by the sensitivity of the isolate. Septicaemia should be treated by parenteral antibiotics. Often treatment with one antibiotic will suffice, but if the patient is neutropenic with a white blood count of <1000 cells/mm^3, a combination of an aminoglycoside and a β-lactam is desirable. Cystic fibrosis patients infected with Pseudomonas in the lungs have, in addition to parenteral antibiotics, also been treated with nebulised aminoglycosides. (Note that several of the organisms have β-lactamases which may inactivate both the cephalosporins and the anti-pseudomonal penicillins and there are several plasmid located aminoglycoside modifying enzymes which inactivate different aminoglycosides.)

Because the coliforms and pseudomonads are important causes of environmentally acquired nococomial infection, efforts directed at decreasing the acquisition and transmission of these organisms within hospitals are important for controlling the rate of hospital acquired infection. In most hospitals in Western Europe the nosocomial infection rate is 8–15%. Monitoring the incidence of infection by recording the occurrence of for example *Pseudomonas* species in high risk areas such as intensive care units is important for infection control. Isolation of patients infected with antibiotic resistant organisms and proper attention to basic hygiene precautions, e.g. handwashing, are similarly important control of infection measures. The use of oral non-absorbable antibiotics active against coliforms and pseudomonads, yet preserving the remainder of the patient's own microflora, (called selective gut decontamination), in order to reduce the incidence of septicaemia and mortality, is at present a controversial topic.

Melioidosis is a common infection in Southeast Asia and the occasional imported case occurs in Europe. The optimal treatment seems to be ceftazidime. Tetracycline or chloramphenicol have also been used to treat patients successfully.

LABORATORY DIAGNOSIS

Septicaemia and focal infections are diagnosed by taking the appropriate specimen (blood culture, urine, sputum etc.) and isolating the organism by routine culture techniques. Because these organisms can also colonise the skin, respiratory tract and urinary system in catheterised patients, the clinical condition of the patient must also be taken into account when deciding upon the relevance of an isolate. Once isolated in pure culture the biochemical reactions of the organism are determined in order to identify it and its sensitivity to antibiotics is also determined.

Pseudomonas pseudomallei can be isolated by culture from blood or tissue specimens and it is identified by the results of biochemical tests. In addition to culture of *Ps. pseudomallei*, serological tests also exist. However, seropositive individuals may be asymptomatic and a positive result may indicate prior exposure to the organism rather than clinical melioidosis.

26   *Salmonella* Species

Salmonella species are Gram-negative bacilli which are usually motile and can grow under aerobic and anaerobic conditions. Unlike other salmonellae, *Salmonella typhi* does not produce gas from glucose. Acute gastroenteritis is the usual presentation of salmonella infection and is commonly caused by *Salmonella enteritidis* or *Salmonella typhimurium*. *Salmonella typhi* and *Salmonella paratyphi* cause enteric fever. Chronic carriage of the organism can develop after the acute infection.

TRANSMISSION   Salmonella infection is usually acquired from contaminated food or water or contact with infected faeces. Ingestion of infected eggs or poultry is the most common source of gastroenteritis but contaminated water supplies are the usual source of enteric fever. The incubation period for gastroenteritis is 6–48 hours and for enteric fever 7–21 days.

CLINICAL FEATURES   Gastroenteritis presents as diarrhoea lasting up to 1 week accompanied by nausea and vomiting. Bacteraemic infection causes a persistent fever. Enteric fever begins with headache, fever and anorexia, diarrhoea only developing in the second week. Rose spots may appear on the upper abdomen.

COMPLICATIONS   Acute gastroenteritis can cause hypovolaemic shock but chronic carriage is rare. Typhoid can cause ulceration and perforation of the ileum and caecum. Between 1 and 3% become chronic carriers. Abscesses can develop in any organ after bacteraemic infection. Osteomyelitis may occur in sickle cell disease.

THERAPY AND PROPHYLAXIS   Gastroenteritis is treated by fluid and electrolyte replacement. Antibiotics, usually ciprofloxacin, are only used in bacteraemia and enteric fever and for chronic carriers. Prevention requires high standards of food hygiene, complete thawing of frozen foods and emphasis on hand washing. Enteric fever is prevented by proper sewage disposal, source isolation of cases and treatment of carriers. Parenteral and oral vaccines are available.

LABORATORY DIAGNOSIS   Gastroenteritis is usually diagnosed by stool culture but blood culture is the primary investigation in bacteraemic infection and enteric fever. *Salmonella typhi* is only excreted in the stool in large numbers in the second week of the disease. Presumptive identification of likely colonies on agar can be made by using specific antisera but this must be confirmed by biochemical tests. Detection of antibodies in the patient's serum is not helpful.

CLINICAL FEATURES

Acute gastroenteritis develops 6 to 48 hours after ingesting the organisms in food or water. Diarrhoea is the common symptom, usually without blood, accompanied by nausea, vomiting, myalgia, headache and a fever lasting up to 2 days. Abdominal cramps occur in most patients and abdominal pain can be mistaken for appendicitis. The diarrhoea usually subsides within one week. Bacteraemia can develop without gastroenteritis, presenting as a persistent fever, due for example to *Salmonella choleraesuis, Salmonella virchow* and *Salmonella dublin*.

Enteric fever is caused by *Salmonella typhi* (typhoid fever) and *Salmonella paratyphi* (paratyphoid fever), the latter being a milder disease. The incubation period is between 7 and 21 days (usually 10–14 days), during which a primary bacteraemia occurs, ending with localisation in the liver and spleen. A secondary bacteraemia coincides with the early symptoms: headache, fever, anorexia, cough, malaise and myalgia. Constipation is common during the first week, followed by diarrhoea in the second week, and there may be a relative bradycardia. Erythematous macules (rose spots) may appear transiently on the upper abdomen. Cervical lymphadenopathy and hepato-splenomegaly may develop. The white cell count is initially raised but falls as the patient becomes anaemic. Without treatment, the disease resolves slowly over 3–4 weeks.

COMPLICATIONS

Diarrhoea can be watery and profuse or bloody and can lead to hypovolaemic shock. Approximately 5–15% of patients excrete the organisms for 1–2 months but very few still do after 6 months.

Myocarditis, hepatic failure and bone marrow depression can occur in severe cases of typhoid fever. Bone marrow aplasia can complicate treatment with chloramphenicol. In the ileum and caecum, lymphoid hyperplasia develops and becomes necrotic after 7–10 days, sloughing to leave ulcers. Most ulcers heal but haemorrhage or perforation of the bowel can occur in 1–2% of patients. Pneumonia in 1–8% of patients is generally a secondary infection. Between 1 and 3% of patients become chronic carriers, particularly those with biliary disease.

Infection can develop in any organ after bacteraemia: meningitis, pneumonia, endocarditis, arteritis, or abscesses of liver, spleen, kidneys and soft tissue. Osteomyelitis is a common complication in patients with sickle cell disease. A reactive arthritis occurs in patients with the HLA-B27 antigen about 10 days after acute infection.

THE BACTERIA

Salmonella species are Gram-negative bacilli which are usually motile and have fimbriae. Cell wall lipopolysaccharide, adhesive factors and toxin production and the host's resistance to colonisation are probably important in pathogenesis. Between 10^6 and 10^9 organisms must be ingested to produce clinical disease except with *S. typhi* in which case very few cells may suffice.

Salmonellae grow readily under aerobic or anaerobic conditions and can survive in water or soil for long periods. Most produce acid and gas from glucose, maltose, mannitol and sorbitol but not from sucrose or lactose. They do not deaminate phenylalanine, produce indole, or hydrolyse urea. They usually form H_2S on triple sugar iron agar and use citrate as sole carbon source. Gas is not formed by *Salmonella typhi*.

Three antigens are important in defining the many serological types of *Salmonella* species. The O antigens are heat stable polysaccharides that form part of the cell wall. The side chains of sugars, but not the core structure, are usually specific. The H antigens are heat-labile flagellar proteins that may show diphasic variation. Bacteria can be changed from one phase to the other by growth in the presence of antiserum. Finally there are surface antigens, e.g. Vi on *S. typhi*, which can inhibit agglutination by O antisera. Serological cross-reactions between salmonellae and enterobacteriaceae are common. In the Kauffmann–White scheme, the main division is into 46 O serogroups. (*S. paratyphi* A is O2, *S. paratyphi* B and *S. typhimurimum* are O4, *S. choleraesuis*, *S. paratyphi* C and *S. virchow* are O7, *S. typhi*, *S. enteritidis* and *S. dublin* are O9). Rough mutants lack the specific side chains. In tracing the spread of an outbreak, phage typing is often used to distinguish strains.

EPIDEMIOLOGY

Infection is usually acquired from contaminated food or water but, among children, can be transmitted directly from faeces. Hospital outbreaks are often associated with person-to-person transmission. Children are most frequently affected and most cases occur at home or in institutions. *Salmonella* species infect poultry, domestic and farm animals and birds but *Salmonella typhi* only infects man. Poultry and eggs are the most important source of human infection. The organisms may contaminate the egg shell from hen faeces or the yolk by haematogenous spread. In the UK, *Salmonella enteritidis*, particularly phage type 4, has become predominant by its spread through chicken flocks. Bulk foods containing raw or uncooked eggs have been associated with large outbreaks. *Salmonella typhimurium* is the most common species causing enteritis in the rest of the world.

Transmission of *Salmonella typhi* occurs by contaminated food or water or direct contact with an infected person or carrier. Large outbreaks are associated with poor sanitation and contamination of the water supply or consumption of shellfish. Typhoid is common in Asia, Africa and South America.

THERAPY AND PROPHYLAXIS

Most cases of gastroenteritis should be treated with fluid and electrolyte replacement but not antibiotics. Patients with bacteraemia or enteric fever should be treated with ciprofloxacin. Chloramphenicol, co-trimoxazole and ampicillin are alternatives but plasmid-mediated resistance to many antibiotics has been encouraged by their use to promote growth in animals. Aminoglycoside resistance of *Salmonella typhimurium* has been increasing in the last 10 years and is present in a quarter of bovine strains. In *Salmonella typhi*, resistance is increasing amongst imported strains especially those from South America, India and Southeast Asia. Ampicillin with probenecid, co-trimoxazole or ciprofloxacin can be used to treat chronic carriers but relapse is common and cholecystectomy may be required.

The animal reservoir can be reduced by slaughter of affected flocks. Cross-contamination between cooked food and uncooked food should be prevented and standards of food hygiene kept high. Early preparation and storage at high ambient temperature, failure to thaw frozen foods completely and inadequate cooking, or reheating, should be avoided. Food poisoning is a notifiable disease and patients admitted with diarrhoea must be placed in source isolation. Hand washing is particularly important in preventing person-to-person spread.

For *Salmonella typhi*, maintenance of an uncontaminated water supply, proper sewage disposal and treatment of long-term carriers is needed. Enteric fever is notifiable and the patient should be source isolated. Three typhoid vaccines are available. The whole cell vaccine is given in two doses and is associated with local and systemic reactions. The polysaccharide vaccine is given in only one dose but can still cause local pain and swelling. The oral live attenuated vaccine has similar efficacy (70–80% protection for 3 years or more) but fewer adverse effects.

LABORATORY DIAGNOSIS

Stool culture is the most important investigation in cases of salmonella gastroenteritis. Blood culture is usually negative but should be performed in patients requiring admission. Enteric fever is diagnosed by isolation of organisms from the blood and sometimes urine. Stool cultures are usually positive in the second week of illness but many samples may be needed to detect chronic carriers. In those given antibiotics, bone marrow culture may be helpful.

Stool specimens are seeded onto selective media (MacConkey agar, xylose lysine desoxycholate agar, desoxycholate citrate agar) and into selenite broth as an enrichment medium. Non-lactose fermenting colonies are investigated as possible salmonellae. Selenite broths are subcultured onto selective media on the second day of incubation. Probable isolates are emulsified on a slide and mixed with polyvalent sera and then sera prepared against specific O antigens to examine for agglutination. Identification must be confirmed by biochemical tests before diagnosis is made, except to ensure that in-patients are source isolated.

Diagnosis of typhoid used to be made serologically by a fourfold rise in the titre of agglutinins to the O antigen (Widal test) but this is not reliable. Less than 50% of untreated patients show such a rise; rises can occur with any febrile illness and results are not interpretable if the patient has been vaccinated against typhoid.

27 *Shigella* Species

Shigella species are non-motile Gram-negative bacilli which can grow aerobically or anaerobically. With a few exceptions, they do not ferment lactose. They cause bacillary dysentery, characterised by the presence of blood and mucus in the stools. There are four species: *Shigella dysenteriae, Shigella boydii, Shigella flexneri* and *Shigella sonnei*, in decreasing order of pathogenicity. *Shigella sonnei* is the most common cause of dysentery in the United Kingdom.

TRANSMISSION   Shigellae are transmitted by the faecal–oral route, on hands, or by contaminated water or food. In tropical countries flies are an important vector. The incubation period is usually 48 hours.

CLINICAL FEATURES   Shigellae cause bacillary dysentery: fever, abdominal cramps and diarrhoea, which may become bloody and mucoid. The severity of illness is highly variable but severe toxaemia is often associated with *Shigella dysenteriae* serogroup 1.

COMPLICATIONS   Dehydration is a cause of infant death in developing countries. Septicaemia and febrile convulsions can occur. Strains producing Shiga toxin are implicated in haemolytic–uraemic syndrome.

THERAPY AND PROPHYLAXIS   Rehydration can usually be achieved by oral fluids. Antibiotic treatment shortens the illness but is limited to more severe cases because bacterial resistance develops readily. Antibiotic-resistant strains are common and ciprofloxacin is probably the drug of choice if antibiotic susceptibilities are not known. Clean water and good sewage disposal are important. Handwashing, disinfection of toilet areas, and proper food handling are useful in outbreak control.

LABORATORY DIAGNOSIS   Stool culture is the usual means of diagnosis. Culture should not be delayed as survival of the organisms in stool is poor. Biochemical identification is required for diagnosis but slide agglutination can give a rapid provisional indication.

CLINICAL FEATURES

After the initial symptoms of fever and abdominal cramps, the patient develops copious watery stools. As the fever resolves, the stool frequency increases and volumes decrease. Stools become bloody and mucoid in half the cases and are accompanied by tenesmus. There is abdominal tenderness and bowel sounds are loud. The rectal mucosa is friable and hyperaemic and coated with mucus.

Symptoms vary from simple diarrhoea to profound toxaemia and last on average for 1 week. Infections caused by *Shigella sonnei* are usually mild and short lived. *Shigella flexneri* causes a more severe illness. *Shigella boydii* and *Shigella dysenteriae* cause disease of variable severity. Strains of *Shigella dysenteriae* serogroup 1 are known as Shiga's bacillus and, in tropical countries, can cause severe abdominal symptoms, diarrhoea and toxaemia. Bacteraemia can occur in 8% of patients, usually children, and recovery is slow.

COMPLICATIONS

Diarrhoea can cause severe dehydration, particularly in the young. It is a leading contributor to infant death in the developing world. Febrile seizures, septicaemia, arthritis and Gram-negative pneumonia may occur. In children, haemolytic anaemia, thrombocytopenia, and renal failure may develop in severe infections caused by toxin-producing strains. Conjunctivitis and iritis are reported in the second week of illness. Excretion of the bacilli usually lasts only a few weeks but can persist for months after dysentery.

THE BACTERIA

Shigella species are non-motile Gram-negative bacilli similar to other Enterobacteriaceae. *Shigella dysenteriae* serogroup 1 produces a cytotoxin, Shiga toxin, which inhibits protein synthesis in tissue culture and causes fluid accumulation in an isolated rabbit ileal loop. It is not known if this is implicated in the greater pathogenicity of this strain. Virulent shigellae can penetrate and multiply within cells in tissue culture.

Shigellae are aerobes and facultative anaerobes and they do not usually ferment lactose or sucrose. They produce colourless colonies on MacConkey agar and desoxycholate citrate agar. Colonies are usually circular and convex but those of *Shigella sonnei* can produce small projections or be coarse and crenated. They do not hydrolyse urea, deaminate phenylalanine, produce lysine decarboxylase or form H_2S on triple sugar iron agar. They fail to grow on citrate medium. They produce acid from glucose and only a few serotypes produce gas.

Shigella dysenteriae comprises 12 serotypes that do not ferment mannitol. The six serotypes of *Shigella flexneri* and the 18 serotypes of *Shigella boydii* ferment manitol but are antigenically distinct. *Shigella sonnei* is a single serotype which ferments mannitol and unlike other shigellae can ferment lactose after 3–5 days.

EPIDEMIOLOGY

Shigellae are pathogens of man and primates. Children between 6 months and 10 years of age are most commonly affected and the disease is more severe in the malnourished. In developing countries, *Shigella flexneri* and *Shigella dysenteriae* predominate and epidemics occur in the summer, flies being an important vector. In the industrialised world, *Shigella sonnei* is the most common and is transmitted by the faecal–oral route, usually within primary schools and nurseries, as the result of inadequate hand washing. Epidemics of shigellosis can occur through contaminated water supplies or food. Once a case has been diagnosed in a household, there is a 40% chance of further cases in children aged 1–4 years in the same house. The rate increases if sanitation is inadequate. Carriage of the organism usually lasts 1–4 weeks but long-term excretion is rare. Acute cases are the most infectious.

THERAPY AND PROPHYLAXIS

Dehydration can usually be corrected by oral fluid intake but intravenous fluids may be necessary in the very young or elderly. Antibiotic treatment will shorten the length of illness and duration of excretion of the organism but is often restricted to the more severe cases because the illness is self-limited and bacterial resistance develops easily. Treatment may be necessary to reduce the infectivity of the acute case. If the organisms are known to be susceptible, ampicillin or co-trimoxazole are effective but, for empirical treatment, ciprofloxacin is most likely to be active. Travellers from Southeast Asia, Africa and South America commonly have trimethoprim-resistant strains. In the UK, *Shigella sonnei* is usually resistant to ampicillin, sulphonamide and tetracycline.

Bacillary dysentery is a notifiable disease and acute cases must be source isolated if in hospital. The patient and carers must be encouraged to wash their hands. Sufferers should be withdrawn from work or food handling until stools are formed. In nurseries, affected children should be excluded as soon as possible and toilet areas disinfected. Adequate cooking and refrigeration of foods is needed. A safe chlorinated water supply and efficient sewage disposal are the most important public health measures in affected areas. Breast feeding is protective for infants. Flies can be reduced by insecticides and removal of refuse. Immunity following an attack is mediated by the IgA of the intestinal mucosa rather than by serum antibodies. Vaccine development is directed at orally administered live strains but no vaccine is yet generally available.

LABORATORY DIAGNOSIS

Faeces are the most important samples for diagnosis. Shigellae do not survive well in faeces and samples should be taken to the laboratory and plated as quickly as possible. Glycerol saline solution can prolong survival. Microscopy of the mucus shows numerous polymorphonuclear leucocytes and red cells but is no more predictive of the diagnosis than a history of bloody diarrhoea. Culture is made on desoxycholate citrate agar or on xylose lysine desoxycholate medium. Likely colonies must be identified by biochemical tests before making a diagnosis but rapid slide agglutination using polyvalent and specific antisera can be useful. Serological diagnosis is rarely made and is of use only in defining outbreaks.

28   *Escherichia coli*

Escherichia coli is the type species of a genus of Gram-negative bacteria which are usually motile and ferment lactose with the production of acid and gas. *Esch. coli* is a common commensal of the human intestine from the neonatal period onwards. However, some strains are capable of causing intestinal disease and others may spread to distant sites and cause infection (commonly urinary tract infection, also septicaemia and meningitis).

TRANSMISSION   Pathogenic strains of *Esch. coli* are frequently transmitted in contaminated food and water but cross-infection in hospital and nursery schools via attendants is common in the very young. In urinary tract infection and septicaemia the causative organism spreads directly from the patient's own bowel flora via the perineum. Disease in neonates is caused by organisms derived from the maternal intestinal flora.

CLINICAL FEATURES   The enteritis caused by *Esch. coli* can vary from a mild diarrhoea to severe diarrhoea with blood and mucus in the stools. In the urinary tract *Esch. coli* usually causes an uncomplicated cystitis with frequency and dysuria but may cause pyelonephritis. *Esch. coli* is a common cause of Gram-negative septicaemia and neonatal meningitis.

COMPLICATIONS   Dehydration may occur, particularly in infants, and this is an important cause of death in developing countries. The haemolytic uraemic syndrome may follow *Esch. coli* infection, specially in children. In the neonate infection may cause meningitis or septicaemia.

THERAPY AND PROPHYLAXIS   Restoration of fluid and electrolyte balance is the most important aspect of treatment of enteric infections. Infections with pathogenic *Esch. coli* can be avoided when travelling abroad by taking care over the food and water consumed. Appropriate cross-infection measures should be taken when cases occur in hospitals or nurseries. Community-acquired urinary tract infection is often susceptible to amoxycillin or co-trimoxazole. Urinary tract infections or invasive disease acquired in hospital should be treated with appropriate antibiotics following sensitivity testing.

LABORATORY DIAGNOSIS   In urinary tract infections a mid-stream specimen of urine (MSU) should be cultured semi-quantitatively. In enteric disease stools can be cultured but pathogenic strains must be identified by the detection of a variety of toxins by either serological or biological methods.

CLINICAL FEATURES

Diarrhoeal Disease

Certain strains of *Esch. coli* elaborate toxins which cause diarrhoea. Some have a cholera-like action so the illness is characterised by watery diarrhoea, cramps, nausea and sometimes low-grade fever. Other toxins lead to mucosal destruction or invasion so that there are dysentery-like symptoms with blood in the stools which may be extensive in haemorrhagic colitis. The incubation period of these diarrhoeal illnesses is 1–2 days and the duration of disease is usually 2–5 days.

Urinary Tract Infection

The infection is usually of the lower urinary tract (cystitis) and presents with sudden onset of frequency, dysuria and urgency of micturition. There may also be some haematuria. Infection of the upper urinary tract (pyelonephritis) is often associated with fever and loin pain. Repeated urinary tract infections may be associated with abnormalities of the renal pelvis, ureters or bladder or the presence of stones. Some types of renal stones are believed to form around a nidus of infection. Repeated infections of the kidney may lead to scarring and impairment of function which may lead to renal failure.

Septicaemia

Esch. coli is the most common cause of Gram-negative septicaemia and is often associated with impaired host defences especially neutropenia. There is fever, low blood pressure with impaired perfusion of the peripheral vasculature and eventually the major organs. The mortality is high (20–30%).

Meningitis

Esch. coli causes meningitis most frequently in neonates in whom the symptoms and signs may be very non-specific, i.e. fever, lethargy, poor feeding, vomiting or diarrhoea. The mortality is 20–30%.

The Haemolytic Uraemic Syndrome (HUS)

HUS may complicate infection with strains of *Esch. coli* which produce verotoxins. The disease is characterised by acute renal failure, anaemia and thrombocytopenia.

THE BACTERIA

Esch. coli is the type species of the genus *Escherichia* within the family Enterobacteriaceae. *Esch. coli* grows on simple media and ferments lactose, a property used in preliminary identification. Strains can be typed serologically on the basis of variations in O (somatic), H (flagellar) and K (capsular) antigens. This property is useful in epidemiological studies and some K serotypes are associated with virulence. However, diarrhoeal disease is associated with the ability of strains to produce toxins and this does not correlate well with serotype. Enterotoxigenic *Esch. coli* produce heat stable and heat labile exotoxins which have a cholera toxin-like action. Enterohaemorrhagic *Esch. coli* produce a different exotoxin which damages vero cells (cultured cells derived from monkey kidney) in culture and hence is called verotoxin. Enteroinvasive *Esch. coli* are those with an ability to invade and destroy colonic epithelial cells by an unknown mechanism. Enteropathogenic *Esch. coli* have pili which mediate adherence to epithelial cells with subsequent destruction of microvilli.

EPIDEMIOLOGY

Esch. coli is an ubiquitous commensal of the gastrointestinal tract where it is present in large numbers. Urinary tract infection and septicaemia are due to endogenous strains and neonatal meningitis is caused by strains from the maternal flora.

Strains which cause diarrhoea are frequently spread from person to person and sometimes via food, milk or water. Enterotoxigenic *Esch. coli* are the most important cause of diarrhoea in infants in developing countries and the commonest cause of traveller's diarrhoea. Water may be an important vehicle for infection. Enteroinvasive and enterohaemorrhagic *Esch. coli* cause outbreaks and sporadic cases worldwide and food and unpasteurised milk are important in their spread. Enteropathogenic *Esch.*

coli cause infection in babies and young children, outbreaks appearing to be due to person-to-person spread.

THERAPY AND PROPHYLAXIS

Diarrhoeal disease due to *Esch. coli* does not require specific antibiotic therapy but in severe cases restoration of fluid and electrolyte balance is important. High standards of hygiene with clean water supplies lead to a reduction in *Esch. coli* diarrhoeal disease.

Urinary tract infection should be treated with antibiotics and many, e.g. amoxycillin or trimethoprim, have proved effective. Such antibiotics may be given without awaiting the results of microbiological tests for infections acquired outside hospital. Those acquired in hospital will more often be due to antibiotic-resistant organisms and therapy should therefore be guided by sensitivity testing. Patients with recurrent urinary tract infection should be investigated for abnormalities of the urinary tract. High standards of catheter care are required to prevent infection in catherised patients.

Cases of *Esch. coli* septicaemia and meningitis should receive systemic treatment often with a combination of antibiotics.

LABORATORY DIAGNOSIS

Esch. coli can be recovered easily from appropriate specimens (MSU, faeces, blood, CSF) by culture on lactose-containing media on which characteristic pigmented colonies appear (colour depends on the indicator used).

Antigenic typing using specific antisera can be performed and sometimes assists epidemiological studies. In diarrhoeal disease somewhat complex biological, serological or genetic tests are required if identification of strains pathogenic for the intestine is required.

29 Vibrios

Vibrios are curved Gram-negative bacilli looking like commas, but they may be difficult to distinguish from enterobacteria under the microscope. *Vibrio cholerae* is a human pathogen causing the disease cholera. *Vibrio parahaemolyticus* may be isolated from shellfish in warm waters and may cause food poisoning in man.

TRANSMISSION Cholera is spread by contamination of water supplies containing faecal material but may be contracted through vegetables and other foodstuffs which have been exposed to contaminated water. The incubation period is around 1–5 days and the infectivity from 5 to 15 days. *Vibrio parahaemolyticus* gastroenteritis arises mostly from food poisoning from shellfish and other marine creatures eaten undercooked or raw, with an incubation period of around 12–15 hours. A major feature of these infections is diarrhoea, in some cases profuse.

SYMPTOMS AND SIGNS Cholera is characterised by the production of copious, watery diarrhoea with dehydration and severe prostration. Infection due to *V. parahaemolyticus* causes relatively mild vomiting and diarrhoea lasting about 3 days.

COMPLICATIONS The major complication of cholera is potentially fatal dehydration.

THERAPY AND PROPHYLAXIS The main treatment of cholera is by rehydration and salt replacement. Antibiotics have a secondary role but tetracycline may shorten the period of diarrhoea and the excretion of the organism. Two doses of a killed *V. cholerae* vaccine confer short-lived immunity lasting up to 6 months.

LABORATORY DIAGNOSIS *V. cholerae* and *V. parahaemolyticus* may be isolated from faeces by the use of the selective medium, thiosulphate citrate bile salt sucrose agar (TCBS medium). Alkaline peptone water enrichment culture may also be used for the isolation of *V. cholerae*.

CLINICAL FEATURES

Classical cholera produces a diarrhoea which is initially faecal in nature, but rapidly gives way to 'rice water stools' which are clear and watery and are flecked with mucus. This diarrhoea may be so copious as to profoundly dehydrate the patient. That picture, most commonly seen with classical cholera, has tended to give way to a more insidious course of a less florid nature caused by the El Tor vibrio which also gives rise to diarrhoea but not commonly to life-threatening dehydration. *V. parahaemolyticus* infection gives rise to combined vomiting and diarrhoea of usually no more than 3 days' duration and not of profound degree.

Classical cholera has been associated with the production of vast amounts of watery stools, leading to progressive and rapid dehydration and salt depletion, culminating in anuria and shock, before death in a proportion of cases.

THE BACTERIA

V. cholerae grows well in enriched alkaline peptone water in 4 to 6 hours. On TCBS medium it grows as large round smooth shiny yellow colonies after overnight incubation. *V. parahaemolyticus* produces large and distinctive green blue colonies with a shiny surface on TCBS medium. The vibrios are very motile in wet preparations under the microscope. Classical *V. cholerae* may be distinguished from the El Tor biotype or from *V. parahaemolyticus* by a variety of laboratory tests involving differential haemagglutination of chick red cells, polymyxin B susceptibility, and specific bacteriophage susceptibility.

EPIDEMIOLOGY

Classical *V. cholerae* still exists endemically in India, Bangladesh and some other Southeast Asian countries, coexisting with the El Tor biotype. The latter has become widespread through many countries in Asia, Africa and South America. Human infection occurs as a result of ingestion of contaminated drinking water into which infected faecal material has permeated. Some cases are acquired as a result of ingestion of foodstuffs such as vegetables which have been washed in contaminated water. Cases are from time to time received in developed countries as a result of air travel, but the disease is highly unlikely to spread in countries with satisfactory sewage disposal arrangements and guaranteed clean drinking water supplies. In the nineteenth century pandemics of classical cholera occurred. The organism produces a potent enterotoxin acting on the small bowel epithelial lining, causing copious outpourings of electrolyte-rich intestinal fluid.

V. parahaemolyticus may occur in coastal waters in warm climes and contaminates shellfish which may then be ingested by man. Person-to-person infection is rare.

THERAPY AND PROPHYLAXIS

Cholera in either of its main forms is most importantly treated by rehydration, salt replacement and attention to acid–base balance. The role of antibiotics is secondary. The main antibiotic used has been tetracycline because it curtails somewhat the period of diarrhoea and may reduce the period of excretion of the organism. More recently the 4-quinolones have been used with some success. The clinical severity of *V. parahaemolyticus* food poisoning is not usually sufficient to justify the use of antibiotics, but a wide variety have been prescribed with little impact on the course of the infection. There is a vaccine of limited value for protection of those who go to cholera areas during the rain season when cholera cases are most numerous. The killed vaccine of *V. cholerae* is given by two intradermal or subcutaneous injections a month apart, and gives limited protection for 4 to 6 months. Contacts of patients with cholera may be given prophylactic tetracycline but immunisation is not considered of value.

LABORATORY DIAGNOSIS

The important laboratory sample is faeces. With *V. cholerae* growth will occur rapidly in alkaline peptone water and this may be used as an enrichment culture. Both *V. cholerae* and *V. parahaemolyticus* may be isolated on the selective medium, TCBS agar.

30 Campylobacter Species

The most common isolate from the stools of patients with bacterial food poisoning in the UK, *Campylobacter* species comprise a genus of Gram-negative motile, spirally curved rods which inhabit the gut of domestic farm and other animals and birds, causing abortion in cattle and sheep and enterocolitis in man.

TRANSMISSION Infection in man follows ingestion of the organism on undercooked food, usually meat. Outbreaks have been noted with unpasteurised milk or contaminated water supplies. Cross-contamination from raw food to cooked food is a common method of transmission. A high proportion of chicken carcasses are infected. Similar risks of food poisoning therefore apply to *Campylobacter* species as to *Salmonella* species.

CLINICAL FEATURES The diarrhoeal illness is similar to that caused by *Salmonella* species except that abdominal pains and bloodstained stools probably occur more frequently. After an incubation period of 3–5 days, the patient experiences griping abdominal pain and starts to have diarrhoea. Vomiting is not a prominent feature initially but the patient may have influenza-like symptoms. The stools are liquid and may become watery often with bile and some blood. Diarrhoea may persist for many days or even weeks and excretion of the organism for 3 months has been described.

COMPLICATIONS Patients may develop a reactive arthritis.

THERAPY AND PROPHYLAXIS It is conventional not to treat acute enteritis with antibiotics—they have little effect in acute salmonellosis or shigellosis. However, erythromycin appears to reduce the length of campylobacter illness by an average of one day but by the time culture results are available the patient is usually well on the way to recovery and will not require specific antibiotic therapy. The disease is prevented by cooking food properly.

LABORATORY DIAGNOSIS *Campylobacter* species are grown from stool specimens plated on selective media in a special macroaerophilic environment. Differentiation of species is not useful clinically but may give a clue to the source of infection. Blood cultures may be positive, particularly in immunosuppressed patients.

CLINICAL FEATURES

The incubation period is 3–5 days and there is usually an acute febrile prodrome with influenza-like symptoms (fever, malaise, headache and myalgia). Coincident with this

is the onset of griping abdominal pain and progression to liquid then watery stools with blood. The illness is sometimes severe and then requires hospital care. Babies tend to produce bloody motions and may have no diarrhoea.

Campylobacters cause acute enteritis. Because abdominal pain is a prominent feature, the symptoms may (but should not) be mistaken for acute intra-abdominal surgical conditions such as acute appendicitis. At operation, the jejunum and ileum are found to be generally inflamed and there is mesenteric node enlargement and there may even be genuine appendicitis. For those where colitis is prominent, differentiation from acute non-infectious (ulcerative) colitis is important because the patient must not be given corticosteroids.

Bloody diarrhoeal stools (dysentery) suggest a differential diagnosis of shigella or amoebic dysentery but these conditions are rarely acquired in the UK. Patients with dysentery from abroad may have any one or more of these three pathogens in the stool.

THE BACTERIA

Campylobacters are spiral Gram-negative rods, diameter 0.2–0.5 μm and length 0.5–5 μm. They become coccoid in air. They are oxidase and catalase positive. They have one or more polar flagella and spin so rapidly in a wet preparation as to hide the spiral shape. Although they respire with oxygen and utilise the Krebs cycle, excess oxygen is toxic and a microaerophilic environment is required to ensure growth in the laboratory. This single fact led to the delay in isolating the organism from colitic stools until the 1970s, even though spiral 'vibrios' had been noted in stools as long ago as 1886. When the proper environment had been established, campylobacters were found to be the most common bacterial causes of acute colitis. A useful selective maneouvre is to grow the organism at 42°C, at which temperature the enterobacteriaceae will not grow.

C. jejuni is the major cause of enterocolitis in man and is frequently isolated from poultry. The organism produces cytotoxin and cholera-like toxin. C. coli is isolated from pigs with dysentery and causes slightly less severe colitis in man than C. jejuni. C. laridis is found in the gut of seagulls and has been responsible for sporadic cases and water-borne outbreaks but only accounts for 0.1% of Campylobacter isolates. C. fetus (which causes abortion in sheep) is isolated occasionally from blood cultures from immunodeficient patients.

EPIDEMIOLOGY

Man acquires campylobacters by eating contaminated meat or drinking contaminated water. As these organisms are found in the gut of animals and birds, it is presumed that carcasses become contaminated by gut contents during butchery.

Many chicken pieces in supermarkets grow salmonellae or campylobacters indicating a need for thorough cooking. Outbreaks have occurred from contaminated unpasteurised milk and drinking water. The disease is notifiable under food poisoning regulations. Documented case-to-case spread seems to be rare. Most outbreaks are due to the improper storage, handling and cooking of food but because of the long incubation period, the responsible food is rarely identified except in outbreaks. Unpasteurised milk and water supplies have also been associated with outbreaks.

THERAPY

In most cases, the disease is self-limiting and no specific antibiotic treatment is indicated. Controlled trials suggest that erythromycin can reduce the length of the illness by an average of a day. If the disease is classical (with severe griping abdominal pains and dysentery), some would prescribe erythromycin, but often by the time the culture results are back, the patient is much better. Invasive C. fetus infections in immunocompromised patients are usually treated empirically and successfully with cephalosporins or fluoroquinolones. There does not appear to be a long-term carrier state.

LABORATORY DIAGNOSIS

Stools are plated directly onto complex selective solid media containing blood agar with vancomycin and other antibiotics. The plates are incubated at 42°C (for *C. jejuni* and *C. coli* and at 35°C for other species) in a gas jar in microaerophilic conditions. Subcultures can be made on non-selective media. Microscopy reveals characteristic small Gram-negative S-shaped rods. Species can be presumptively distinguished by simple tests (e.g. many strains of *C. jejuni* hydrolyse hippurate and *C. fetus* tends to be resistant to cephalothin), but it is probably not worth doing these tests in the routine laboratory because they are not 100% specific and the species type is of little clinical significance. Full identification would, however, be performed in the investigation of an outbreak.

31 *Helicobacter pylori*

Helicobacter pylori is a slowly growing, microaerophilic Gram-negative organism that is found in the stomach and duodenum and is related to a number of gastroduodenal diseases. It is characterised by the abundant production of the enzyme urease, which is a virulence factor and can be used diagnostically.

TRANSMISSION   The prevalence of infection is much higher in under-developed countries than in industrially developed countries. The transmission of the organism probably occurs from person to person by faeco-oral spread. Drinking water has been suggested as a possible environmental source of infection.

CLINICAL FEATURES   The presence of *H. pylori* in the stomach inevitably leads to inflammation, which is often asymptomatic. The principal illness caused by *H. pylori* are gastric and duodenal ulceration. The role of *H. pylori* in non-ulcer dyspepsia is controversial.

COMPLICATIONS   Colonisation by *H. pylori* is an important risk factor for the development of gastric lymphoma and carcinoma.

THERAPY   Eradication of the organism leads to symptomatic improvement in some patients with non-ulcer dyspepsia and healing of peptic ulceration in the majority of patients. Eradication of the organism is correlated with a low ulcer relapse rate. Currently the most effective regimens are bismuth + metronidazole + tetracycline given for 2 weeks or omeprazole + amoxycillin + metronidazole given for 2 weeks.

LABORATORY DIAGNOSIS   Colonisation of the stomach can be detected in a gastric biopsy specimen by histology, culture, the rapid urease test, *in situ* hybridisation or the polymerase chain reaction. The presence of the organism can also be determined by serology or the urea-breath test.

CLINICAL FEATURES

H. pylori is the cause of a number of gastroduodenal diseases. There is no well defined, acute illness associated with *H. pylori* infection, although hypochlorhydria, nausea and vomiting have been recorded. Infection with *H. pylori* can be asymptomatic but in all cases there is histological evidence of gastric inflammation. The relationship between infection and non-ulcer dyspepsia is controversial. However, there is evidence that eradication of the organism does lead to symptomatic improvement in some patients. There is abundant evidence showing that *H. pylori* is a significant cause of peptic ulceration, with virtually all duodenal and 70% of gastric ulcers causally linked to the

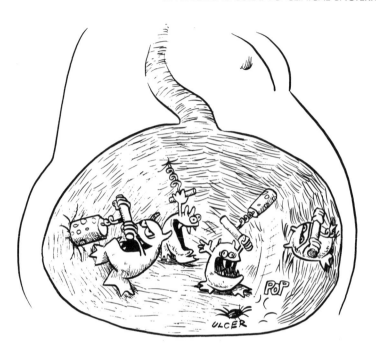

presence of the organism. There is prospective sero-epidemiological evidence showing that infection with *H. pylori* over a long period of time (20–30 years) is a significant risk factor for the development of gastric carcinoma. Populations with a high prevalence of infection have a sixfold increased risk of developing gastric carcinoma compared to populations with a low prevalence of infection. There is also both epidemiological and experimental evidence indicating that *H. pylori* causes gastric beta-cell lymphoma.

The mechanism(s) of the pathogenesis of inflammation and ulceration have not been clearly defined. *H. pylori* binds only to gastric epithelium and its presence leads to both a disruption of the protective mucous layer caused by the bacterial phospholipase A2 and the production of ammonia and an inhibition of mucus secretion from goblet cells by some unknown mechanism. Epithelial cell damage is caused by the ammonia and by the secretion of a cytotoxin from a proportion of the strains. *H. pylori* secretes a chemoattractant and also activates neutrophils thus exacerbating damage to the epithelium. In addition to these direct effects upon the gastric epithelium, some antibodies produced in response to the organism have been found to cross-react with gastric tissue and cause damage. Also the organism disturbs the normal physiological control of acid secretion by inhibiting somatostatin secretion which leads to hypergastrinaemia and therefore excess acid secretion.

H. pylori is principally found in the gastric antrum and in regions of gastric metaplasia in the duodenum leading to a duodenitis and in some patients, duodenal ulceration. *H. pylori* is also found in the fundus of the stomach where it may lead to gastric ulceration. With long-term colonisation and chronic inflammation of the stomach, atrophic gastritis and intestinal metaplasia may supervene which ultimately may lead to the development of gastric carcinoma.

THE BACTERIA

The genus *Helicobacter* comprises a number of species isolated from the gastrointestinal tracts of animals and humans. *H. pylori* was first isolated in 1983 and classified as a novel genus in 1989. *H. pylori* is a Gram-negative, spiral-shaped organism that grows slowly taking 5 days to form visible colonies on bacteriological media. It is microaerophilic and requires CO_2. The organism is found in the stomach and duodenum of humans and rhesus monkeys. Other *Helicobacter* species that have been isolated are *H. felis* (cat and dog stomach); *H. mustelae* (ferret stomach); *H. acinonyx* (cheetah stomach); *H. nemestrinae* (macaque stomach); and *H. cinaedi*, *H. fenelliea* (human faeces). The natural habitat of *Helicobacter* is the sub-mucous layer of the gastrointestinal tract and most of the organisms that inhabit this ecological niche including *Helicobacter*, have three biological characteristics in common, namely a spiral shape, production of urease enzyme and microaerophilism.

H. pylori possesses surface adhesins which bind specifically to receptors on gastric mucosa. The organism produces urease and phospholipase which are important virulence properties. Some strains of *H. pylori* are ulcerogenic, producing a vacuolating cytotoxin; these strains are more likely to be found in patients with duodenal ulceration rather than non-ulcer dyspepsia.

EPIDEMIOLOGY

H. pylori is one of the most common bacterial infections worldwide and occurs equally in males and females. In industrially developed countries with a good public hygiene infrastructure, the seroprevalence increases progressively with age from a prevalence of about 5% (<10 years) to 60% (>50 years). In under-developed countries, there is no age-related difference in seroprevalence, 80% of individuals <5 years old being infected. The age-related increase in developed countries probably reflects the cohort effect of a greater prevalence of infection in society when those who are now 50–60 years old were children.

The mode of transmission of *H. pylori* is not known but epidemiological evidence suggests both a person-to-person spread and infection derived from public water sources. Clustering of infection occurs in families and closed institutions and the method of spread is either oro-oral or faeco-oral. The isolation of *H. pylori* in faeces supports the latter.

THERAPY

The optimal therapeutic regimen for *H. pylori*-associated gastroduodenal disease is not yet established. H2 receptor antagonists will heal ulceration but because the drugs have no effect upon *H. pylori*, as soon as the drugs are stopped, the ulcer recurs at a high frequency. Eradication of the organism results in a very low ulcer recurrence rate. Eradication rates of 95% can be achieved by a combination of a bismuth compound with a nitroimidazole (metronidazole or tinidazole) and either amoxycillin or tetracycline. This combination is usually given for 2–4 weeks. Similar eradication rates, and hence ulcer healing rates, can be achieved with a combination of a proton pump inhibitor (omeprazole, lansaprazole), amoxycillin and metronidazole given for 2 weeks.

Gastric ulceration will respond to the same therapeutic regimen. Controversy still exists about the need to eradicate *H. pylori* from patients with non-ulcer dyspepsia, but a subgroup of patients may derive symptomatic benefit from eradication of the organism. There is as yet no suggestion that eradication should be attempted in asymptomatic carriers, but because of the increased risk of carcinoma, the development of a vaccine is being pursued. Eradication of *H. pylori* can lead to regression of low-grade gastric beta-cell lymphomas.

LABORATORY DIAGNOSIS

Colonisation by *H. pylori* can be detected by a variety of techniques which can be either invasive or non-invasive. Invasive methods require upper gastrointestinal endoscopy with collection of a gastric biopsy. *H. pylori* can be detected in such specimens either by histological examination of a Giemsa-stained section, culture of the specimen on a selective medium, detection of urease enzyme (a surrogate marker for *H. pylori*) by a biopsy urease test or by the use of *in situ* hybridisation or the polymerase chain reaction. The currently accepted 'gold standard' for detection is a combination of culture and histology. Only with isolation by culture can one test the antibiotic sensitivity of the isolates, which may be important in determining the correct treatment, but primary isolation takes 5 days. The most rapid diagnostic test is the biopsy urease test.

Non-invasive techniques for the detection of *H. pylori* are serology and the urease breath test. Serology has been used principally for epidemiological studies but can also be used as a screening procedure to reduce endoscopy workload and to monitor the effectiveness of treatment. The urea-breath test involves giving the patient either ^{13}C- or ^{14}C-labelled urea as a test meal and measuring the excretion of $^{13}CO_2$ or $^{14}CO_2$ in the breath.

32 *Bacteroides* Species

Bacteroides and related organisms form a group of anaerobic, non-sporing, Gram-negative rods which are common commensals of the mouth, the gastrointestinal tract and the lower genital tract in women. They are common causes of certain types of wound infection and abscesses.

TRANSMISSION The majority of infections are endogenous in origin although the organisms are sometimes isolated as part of a mixed flora causing wound infection after animal or human bites.

CLINICAL FEATURES Wound infections and abscesses due to bacteroides are associated with large quantities of foul-smelling pus. Such complications were relatively common after abdominal or pelvic surgery. Related organisms are associated with severe periodontitis. Cerebral abscesses, lung abscesses and aspiration pneumonia may involve bacteroides and related organisms.

COMPLICATIONS *Bacteroides* species are the most common anaerobic bacteria isolated in cases of bacteraemia and septicaemia.

THERAPY AND PROPHYLAXIS Surgical drainage of pus is essential as is surgical debridement when necessary. Metronidazole is the most commonly used antibiotic for both treatment and prophylaxis prior to abdominal and pelvic surgery.

LABORATORY DIAGNOSIS Samples of pus should be sent to the laboratory. A rapid result can be obtained using gas chromatography; otherwise the organisms must be cultured in an anaerobic jar or cabinet.

CLINICAL FEATURES

Bacteroides are most commonly associated with wound infection following abdominal or pelvic surgery. The infection is associated with the production of large amounts of foul-smelling pus. Infections are often associated with necrotic tissue and there may be gas in the tissues. The pus sometimes fluoresces red under ultraviolet light. Infection may lead to abscess formation which is liable to occur in deep sites in the pelvis or abdomen. Low-grade fever, abdominal or pelvic pain and leucocytosis following surgery may indicate the presence of an abscess. A related organism, now called *Porphyromonas gingivalis*, is associated with the marked inflammation, bleeding and gingival loss of severe periodontitis.

Anaerobic flora from the oropharynx are often found in lung abscesses, empyema and aspiration pneumonia.

COMPLICATIONS

Bacteroides infection at any site may lead to bacteraemia which sometimes leads to signs of septic shock.

THE BACTERIA

The organisms are strictly anaerobic, Gram-negative bacilli which ferment sugars and produce a typical pattern of metabolic end products. Recently both the classification and nomenclature of the group have been changed. *Bacteroides* species are still the most frequent anaerobes among the normal faecal flora. Although not the most abundant *Bacteroides* species is the normal flora, *B. fragilis* is the most significant species in terms of wound infection and pelvic and abdominal abscesses.

  B. gingivalis has now been reclassified and named *Porphyromonas gingivalis* and many of the black-pigmented species of *Bacteroides* have been assigned to the genus *Prevotella*.

EPIDEMIOLOGY

Bacteroides and related species are the organisms found in greatest abundance in the normal faecal flora. The mean bacterial count is approximately 10^{10} per gram of faeces, two or three logs more than the aerobic flora. They are also found as a component of the oral and genital tract flora.

As a consequence most human infections are endogenous and wound infection and abscess formation are particularly associated with bowel surgery or gynaecological surgery involving a breach of a mucosal surface.

Bacteroides and related species are sometimes found in the mixed anaerobic flora infecting wounds from human and animal bites.

THERAPY AND PROPHYLAXIS

Drainage of pus and debridement of any necrotic tissue is a most important aspect of the management of bacteroides infections. The organisms are sensitive to a variety of antibiotics but metronidazole is the treatment of choice. It is also given prophylactically, often in combination with, for example, a cephalosporin (to cover possible aerobic organisms) as a single dose or short course peri-operatively in relation to abdominal or pelvic surgery. This has reduced significantly the occurrence of infectious complications post-operatively.

LABORATORY DIAGNOSIS

Samples of pus are preferred to wound swabs for the diagnosis of bacteroides infection. The foul smell of the pus may suggest anaerobes and a Gram-stained smear may be helpful. Specific immunofluorescence test are also available. Alternatively the metabolic products in the pus can be analysed in a gas chromatograph since the pattern of volatile fatty acids produced is characteristic. This permits diagnosis within an hour or two of receipt of the specimen.

Culture of the organism requires inoculation onto suitable media and incubation under strictly anaerobic conditions in an aerobic jar or cabinet. The bacteria can be slow growing requiring incubation for a number of days before they can be identified.

33   The Mycobacteria

Mycobacteria are acid-fast bacilli that stain red with the Ziehl–Neelsen stain. They are aerobic organisms that may take 2–12 weeks to grow on artificial media such as Löwenstein–Jensen medium. *M. leprae* cannot be cultured on artificial media. There are many different species which cause disease ranging from localised skin infections through pulmonary illness to disseminated disease.

TRANSMISSION   *M. leprae* and *M. tuberculosis* are transmitted from an infected patient by nasal secretion in the case of leprosy and droplet infection in the case of tuberculosis. All other mycobacteria are found in the environment and infection is acquired either by skin inoculation or through the gastrointestinal tract from contaminated water or food.

CLINICAL FEATURES   Tuberculosis is caused mainly by *M. tuberculosis*. This presents as a chronic pulmonary infection with fever and weight loss. Dissemination to all locations in the body produces many signs and symptoms depending upon the area most affected. Localised skin lesions are caused by *M. marinum* and ulceration by *M. ulcerans*. Localised skin abscesses are caused by *M. chelonei/M. fortuitum* which also cause a cervical adenitis, as does *M. scrofulaceum* and *M. avium-intracellulare*. *M. avium-intracellulare* is also a cause of disseminated disease in AIDS patients.

   M. leprae is the cause of leprosy which presents with skin and peripheral nerve involvement leading to tissue destruction and deformity due to loss of sensation and motor function.

THERAPY AND PROPHYLAXIS   Tuberculosis is treated with rifampicin, isoniazid, pyrazinamide and ethambutol. Other agents that are used are streptomycin, ethionamide and thiacetazone. Mycobacterial disease other than that caused by *M. tuberculosis* can be treated with minocycline, quinolones, co-trimoxazole and macrolides depending upon the illness. Leprosy is treated with rifampicin, dapsone and clofazimine. Vaccination with BCG protects against both tuberculosis and leprosy.

LABORATORY DIAGNOSIS   Diagnosis of mycobacterial infections relies upon microscopy using the Ziehl–Neelsen stain and culture (except for *M. leprae*) on Löwenstein–Jensen medium. Serology is not useful in diagnosis. Rapid diagnosis of *M. tuberculosis* and *M. avium-intracellulare* can be achieved using DNA amplification techniques.

CLINICAL FEATURES

Tuberculosis is caused by *M. tuberculosis*, although a smaller number of cases are caused by *M. kansasii* and *M. xenopi*. Primary tuberculosis occurs in an individual not

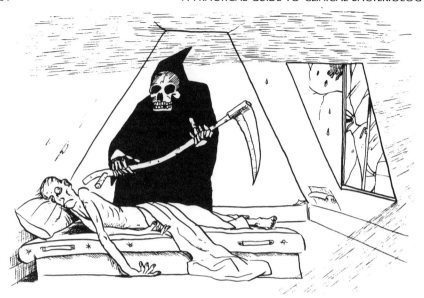

previously exposed to mycobacteria and results in a local infection of the regional lymph nodes. It is usually asymptomatic. In some individuals the primary disease may present as a serious disseminated infection (miliary tuberculosis) with widespread granulomas in many organs. Meningitis may also occur following dissemination to the brain.

Post-primary tuberculosis occurs following reactivation or reinfection. There is enlargement and necrosis of the primary complex, leading to a cavity in the lungs and there may be rupture of the focus into a bronchus leading to the production of infected sputum. The patient presents with weight loss, night sweats, a chronic cough, a low-grade temperature and haemoptysis. There may be a pleural effusion. Because of the dissemination of the organism in the primary infection, tuberculosis may present with a diverse range of local signs and symptoms.

M. avium-intracellulare causes a disseminated infection in patients with AIDS that presents with a low-grade temperature, weight loss, night sweats, diarrhoea and anaemia. *M. marinum* causes localised warty skin lesions and is known as 'swimming pool granuloma' because of the association with contact with water either in swimming pools or fish tanks. The organism may also cause cervical adenitis. *M. ulcerans* causes Buruli ulcer. The infection starts as an itchy papule which eventually ulcerates, the necrosis spreading deep into the subcutaneous tissues to produce an extensive ulcer with undermined edges. *M. chelonei/M. fortuitum* causes post-injection abscesses or post-operative infections, e.g. osteomyelitis of the sternum. Disseminated infections may occur in immunocompromised patients and the organisms are a cause of cervical adenitis.

Leprosy is a disseminated infection principally affecting the peripheral nerves and skin which has two forms. Patients with tuberculoid leprosy have a high degree of immunity to *M. leprae* and very few bacilli are found in the tissues. This form of the disease presents with erythematous patches on the skin with thickened, tender nerves. There is loss of sensation leading to repeated accidental injury which, along with motor dysfunction, leads to deformity. Patients with lepromatous leprosy have a low degree of immunity to *M. leprae* and large numbers of bacilli are found in peripheral nerves and skin. The disease presents with extensive skin involvement with erythematous plaques and nodules. Thickening of the skin of the face occurs, producing the characteristic leonine face.

THE BACTERIA

The genus *Mycobacterium* comprises at least 57 separate species. Mycobacteria are rod-shaped organisms, $0.5 \times 4\,\mu m$ that have a complex, waxy cell wall. The organisms are called acid- and alcohol-fast bacilli because the dye used in the Ziehl–Neelsen stain cannot be removed by either mineral acid or alcohol and so mycobacteria appear red with this staining method. Mycobacteria are mainly obligate aerobic organisms and are cultured principally on Löwenstein–Jensen medium. They grow very slowly and may take from 2 to 12 weeks to form visible colonies on primary culture. Some species produce a bright orange or yellow pigment. *Mycobacterium leprae* cannot be cultured on artificial media.

The main pathogenic species are: *M. tuberculosis; M. kansasii; M. marinum; M. avium-intracellulare; M. ulcerans; M. xenopi; M. chelonei; M. fortuitum* and *M. leprae*. Mycobacteria other than *M. tuberculosis* and *M. leprae* are often called 'atypical' mycobacteria.

The virulence determinants of mycobacteria are poorly understood but the organisms are intracellular pathogens and once phagocytosed are able to prevent phagolysosome fusion. Mycobacteria have profound effects upon the human immune system and are able to subvert the normal protective response and turn it into a tissue destructive response.

EPIDEMIOLOGY

M. tuberculosis is found worldwide and infects about 100 million people. The organism is transmitted by droplet infection and the source of infection is another human case. Overcrowding is a factor which predisposes to cross-infection. The most infectious patients are those whose sputum contains micro-organisms seen on microscopy (smear positive). Cervical adenitis and tuberculosis of the gastrointestinal tract are contracted by drinking unpasteurised milk. This form of tuberculosis is rare in the UK and is usually caused by *M. bovis*, a member of the *M. tuberculosis* complex.

M. leprae is found in tropical and subtropical countries and infects between 10–11 million people. Infection occurs in children or adolescents but because of the long incubation period of 2–10 years, clinical illness usually presents in adults. *M. leprae* is a human pathogen with no other natural reservoir of infection. Borderline lepromatous and lepromatous cases are the most infectious, with large numbers of bacilli in nasal secretions. Transmission occurs in conditions of overcrowding.

Other mycobacterial species causing human disease are environmental organisms and are found in soil and water. Infection caused by these organisms follows ingestion or skin inoculation. The immune competence of the host is often a decisive factor in the development of disease as opposed to colonisation with these opportunistic 'atypical' mycobacteria.

THERAPY AND PROPHYLAXIS

Tuberculosis is treated with a combination of rifampicin, isoniazid and pyrazinamide for 9 months, with pyrazinamide given for the first 2 months only. An alternative regimen is a combination of rifampicin, isoniazid, ethambutol and pyrazinamide given for 6 months with pyrazinamide and ethambutol given for the first 2 months only. A variety of regimens, e.g. amikacin + rifampicin + ethambutol + ciprofloxacin, have been used to treat disseminated *M. avium-intracellulare* infection in AIDS patients because the organism is resistant to the commonly used drugs. *M. kansasii* and *M. xenopi* infections are treated with standard first line agents. *M. marinum* infections are self-limiting but minocycline is an effective drug. *M. ulcerans* infections are treated with surgery and clofazimine and quinolones have been shown to be active. *M. chelonei/M. fortuitum* infections respond to co-trimoxazole, macrolides or quinolones.

Public health measures are important in the prevention of tuberculosis. Pasteurisation of milk and the veterinary surveillance of herds is important in eradication of gastrointestinal disease. Notification of cases of tuberculosis and contact tracing is a vital component of tuberculosis control. In hospitals, infected patients should be isolated for the first 2 weeks of treatment. The other component of tuberculosis control is vaccination using BCG. Vaccination of 10–14 year olds is practised in the UK. BCG vaccination of Mantoux negative children (see below) of any age should occur if they have contact with infected individuals. In the UK vaccination provides about 80% protection. BCG can cause disseminated infection in immunocompromised patients.

The treatment of leprosy depends upon the type of disease. In lepromatous leprosy treatment is with a combination of dapsone, rifampicin and clofazimine for 2 years. In tuberculoid leprosy, 6 months' treatment with rifampicin and dapsone is used. Few patients with leprosy are encountered in the UK and they should be referred to a leprologist for treatment.

LABORATORY DIAGNOSIS

Mycobacteria can be demonstrated in sputum or bronchial lavage specimens from patients with tuberculosis by a Ziehl–Neelsen stain. Mycobacteria can also be detected in tissue sections, cerebrospinal fluid and pus using the Ziehl–Neelsen stain. A more rapid and specific method of detecting mycobacteria in clinical samples is the use of the polymerase chain reaction (PCR) to amplify DNA. Currently PCR can be used to detect *M. tuberculosis* and *M. avium-intracellulare*.

Specimens are cultured on Löwenstein–Jensen medium. Sputum and urine samples should each be pooled before inoculating the medium because of the low numbers of organisms and the unpredictability of excretion. Primary isolation can take 1–12 weeks depending upon the species of mycobacteria. *M. tuberculosis* usually produces visible growth in 2–4 weeks.

The Mantoux test is inoculation of mycobacterial antigens (purified protein derivative) into the skin and assessment of the extent of induration 72 hours later. This test does not distinguish active disease from previous exposure to mycobacteria, including vaccination with BCG. The test is only rarely helpful diagnostically and in the UK is used to determine if individuals require BCG vaccination. Serology is unhelpful in the diagnosis of mycobacterial infections.

Leprosy is a clinical diagnosis. *M. leprae* is detected in smears from sub-epidermal skin sections by staining with the Ziehl–Neelsen stain. The organism cannot be cultured on artificial media. The lepromin skin test is not useful diagnostically but can be used to indicate the immunological status of the patient. The skin test is positive in tuberculoid leprosy and negative in lepromatous leprosy.

34 The Mycoplasmas

The mycoplasmas are the smallest free-living organisms. They are pleomorphic and Gram-negative, and distinguished from true bacteria by the absence of a cell wall and inability to synthesise cell wall precursors. Unlike viruses, they are capable of independent existence in cell-free media. They are ubiquitous saprophytes or parasites of man, other animals and plants, causing a variety of diseases. Two genera are of medical importance, *Mycoplasma* species and *Ureaplasma* species.

TRANSMISSION In humans they are transmitted via aerosol (respiratory infection), and sexually.

CLINICAL FEATURES *M. pneumoniae* is an important cause of primary atypical pneumonia which is characterised by headache, cough, and flu-like symptoms. *M. hominis* and *U. urealyticum* have a role in non-gonococcal urethritis and pelvic inflammatory disease and possibly other genital and urinary tract disease.

COMPLICATIONS *M. pneumoniae* may be associated with extrapulmonary disease, e.g. myocarditis, haemolytic anaemia, arthritis. Infection with genital mycoplasmas may result in infertility or neonatal infection.

THERAPY Mycoplasmas are sensitive to tetracyclines, and macrolides (e.g. erythromycin) except *M. hominis* which is resistant to most macrolides, but unlike *U. urealyticum*, is sensitive to clindamycin. All mycoplasmas are resistant to cell wall active antibiotics, and also to rifampicin, suphonamides and trimethoprim.

LABORATORY DIAGNOSIS Laboratory isolation of *M. pneumoniae* is seldom undertaken routinely. Serum antibody levels are determined either by complement fixation or gelatin particle agglutination tests. Paired sera are preferred to demonstrate rising titres of antibody. The isolation of genital mycoplasmas from the upper genital tract and extra-genital tract sites is likely to be significant. Serology has no role in the routine diagnosis of genital mycoplasma infection.

CLINICAL FEATURES

Respiratory Infection

M. pneumoniae is an important cause of primary atypical pneumonia (i.e. non-responsive to β-lactam antibiotics). Following infection, pneumonia is more likely to occur in children and adolescents than in adults. Malaise and headache may precede pneumonia by several days, and prodromal symptoms are often mild, with auscultatory signs (e.g. crackles) only becoming marked later, usually affecting a lower lobe. Gastrointestinal symptoms (anorexia, nausea, vomiting and diarrhoea),

and joint and muscle pains are common. The disease may run a protracted course, with cough persisting for weeks, and relapse likely to occur. Cold agglutinins with anti-I blood group specificity occur in some 50% of cases, and may be used as a non-specific diagnostic test. Other less common but important extrapulmonary complications include myocarditis and pericarditis, rashes including erythema multiforme and Stevens–Johnson syndrome, arthritis and meningoencephalitis.

Genital Tract Infection

The genital mycoplasmas are common commensals in the genital tract of sexually active persons. Up to 70% of women harbour *U. urealyticum*, and 40% *M. hominis* in the vagina. Somewhat lower figures are quoted for the male urethra. The pathogenic role of these organisms in a particular disease syndrome is consequently not always clear cut. Table 34.1 indicates current opinion on the role of these organisms in genital and association infection.

THE BACTERIA

Mycoplasmas are small (300 nm) Gram-negative pleomorphic organisms. They stain poorly with most stains, Giemsa being an exception. They do not have a rigid cell wall, being bounded by a trilaminar membrane. Some strains are motile, although they do not have pili or flagella. They are the smallest organisms capable of

Table 34.1 Role of mycoplasmas in genital tract and associated infection.

Disease	Evidence for a mycoplasmal aetiology	
	M. hominis	U. urealyticum
Non-gonococcal urethritis	−	+ +
Cervicitis	−	−
Urethral syndrome	−	+
Bacterial vaginosis	+	−
Pelvic inflammatory disease	+ +	−
Pyelonephritis	+ +	−
Epididymitis	+	−
Chronic prostatitis	−	−
Chorioamnionitis	−	+ +
Postpartum/abortion pyrexia	+ +	+
Low birthweight	−	+
Neonatal respiratory disease and meningitis	+	+ +

independent (cell-free) existence, but require highly nutrient culture media for propagation.

EPIDEMIOLOGY

Mycoplasmas are widely distributed in both animals and plants. They are common commensal organisms of the respiratory and genital tract of animals. Tissue cultures are susceptible to mycoplasma contamination, and this may affect the results of diagnostic tests using this approach. Twelve species of mycoplasmas have been isolated from humans mostly from the oropharynx but only *M. pneumoniae*, *M. hominis*, *M. genitalium* and *U. urealyticum* have been unequivocally demonstrated to cause human disease.

THERAPY

Mycoplasmas are sensitive to tetracyclines, although up to 30% of *M. hominis* and 10% of *U. urealyticum* may be resistant. *M. pneumoniae* and *U. urealyticum* are sensitive to erythromycin and other macrolides, whilst *M. hominis* is resistant. In contrast, *M. hominis* is sensitive to clindamycin (a lincosamine), and *U. urealyticum* resistant. This differential sensitivity has important implications in the therapy of genital infections involving mycoplasmas, when tetracyclines cannot be used. Mycoplasmas are resistant to all cell wall active antibiotics, e.g. penicillins and cephalosporins. Folic acid pathway inhibitors (sulphonamides and trimethoprim), and rifamycins are also inactive. Aminoglycosides (e.g. gentamicin) show moderate activity *in vitro*, but are not clinically effective.

LABORATORY DIAGNOSIS

Primary isolation is best performed in liquid or biphasic medium (liquid with a solid agar slope). Enriched media are required, utilising serum and yeast extract. Mycoplasma growth is recognised by incorporating a suitable substrate (e.g. glucose for *M. pneumoniae*, arginine for *M. hominis*, and urea for *U. urealyticum*), and detecting a pH change with a chemical indicator. Isolation is impracticable for the routine diagnosis of *M. pneumoniae* infection, because the yield is generally poor. For specimens from the genital tract, a semi-quantitative technique is preferred, to give some indication of the mycoplasma load.

Measurement of serum antibody levels is not helpful in genital mycoplasma infections but remains the most reliable method of diagnosing *M. pneumoniae* infection. Complement fixation or gelatin particle agglutination tests are used. The latter detects both IgG and IgM, and therefore is of particular use in the early diagnosis of mycoplasma pneumonia. However, as the test does not distinguish between the immunoglobulin classes, paired sera must be examined to demonstrate a fourfold rise in titre.

35 The Chlamydiae

Chlamydiae are Gram-negative cocci, which have a unique and well-defined life cycle in the cytoplasm of infected cells. They are energy parasites, and cannot be grown on artificial culture media. They are widespread in man and other animals, causing a variety of clinical syndromes. Three species are currently recognised: *Chlamydia psittaci* (a zoonosis sometimes infecting man), *C. trachomatis* (oculo-genital infection almost exclusively affecting man), and *C. pneumoniae* (human respiratory infection).

TRANSMISSION   They are transmitted via aerosol (respiratory infection), inoculation (conjunctivitis) and sexually.

CLINICAL FEATURES   *C. psittaci* and *C. pneumoniae* cause atypical pneumonia and the latter is also responsible for upper respiratory tract illnesses. Various serotypes of *C. trachomatis* cause ocular disease ranging from acute conjunctivitis to trachoma. Other serotypes cause the genital infections, non-gonococcal urethritis and lymphogranuloma venereum.

COMPLICATIONS   *C. psittaci* is a cause of endocarditis. Trachoma may lead to blindness. Genital infection may lead to infertility, strictures or fistulas.

THERAPY AND PROPHYLAXIS   Chlamydiae are sensitive to tetracyclines, macrolides (e.g. erythromycin), and newer fluoroquinolones (e.g. ofloxacin). Zoonotic infection can be prevented by quarantine. Improved ocular hygiene can prevent conjunctivitis and sexual transmission can be prevented by the use of barrier contraceptives.

LABORATORY DIAGNOSIS   Cell cultures are required for the isolation of chlamydiae. Routine diagnosis is made by direct examination of clinical material for antigen using polyclonal or monoclonal antibodies and fluorescence or enzyme-linked immunosorbent assays. Serum antibody assays are of limited use in acute chlamydial infections.

CLINICAL FEATURES

Respiratory Infection

Psittacosis in man presents as an atypical pneumonia characterised by cough, headache and high fever. Endocarditis and severe toxaemia and death are rare complications. It is likely that many cases of psittacosis reported previously are in fact *C. pneumoniae* infections. *C. pneumoniae* is now thought to be one of the commonest bacterial causes of human respiratory infection, affecting predominantly children and young adults. The clinical picture ranges from pharyngitis or laryngitis to pneumonia, characterised by hoarseness and prolonged unproductive cough. Infection may either be sporadic or

occur in outbreaks. There is some evidence of an association between *C. pneumoniae* infection and the onset of wheezing and asthma.

In the newborn *C. trachomatis* is an important cause of respiratory infection with or without conjunctivitis. Infection is acquired at birth, and respiratory symptoms appear at 3 weeks to 3 months of age.

Trachoma

C. trachomatis serovars A to C cause a follicular conjunctivitis which may progress via repeated infection to severe conjunctival scarring (cicatrisation) and blindness. Trachoma is endemic in many parts of the developing world, and an estimated 400–500 million people are infected, with many millions blind. Transmission occurs in association with poor hygiene via direct contact inoculation, and flies.

Paratrachoma

Ocular infection with the oculo-genital serovars of *C. trachomatis* (D to K) also leads to an acute follicular conjunctivitis. Untreated, the disease runs a fluctuating course, before healing without cicatrisation. Corneal clouding may persist. This infection is invariably associated with genital infection of the patient or their partner. Infection of the neonate at birth leads to chlamydial ophthalmia neonatorum.

Genital Tract Infection

C. trachomatis serovars D to K are usually sexually transmitted. In men, the organism is a major cause of non-gonococcal urethritis (NGU), being identified in some 30% of cases. Post-gonococcal urethritis is also commonly chlamydial, resulting from concurrent infection with *Neisseria gonorrhoeae* and *C. trachomatis*. Epididymitis is a recognised complication of chlamydial urethritis, but a direct link with male infertility is lacking. Infection of the female genital tract may cause urethritis, cervicitis, endometritis, salpingitis and extend into the peritoneal cavity (perihepatitis), but inapparent infection is common. Salpingitis is often of polymicrobial aetiology, but chlamydiae are implicated in 30 to 40% of cases. Post-infection tubal damage is a common cause of ectopic pregnancy and infertility.

Lymphogranuloma venereum (LGV) is a systemic infection caused by *C. trachomatis* serovars L1, L2 and L3. The initial genital lesion is a small vesicle or ulcer. On healing, the inguinal lymph nodes enlarge and break down discharging pus. Healing occurs with fibrosis and impairment of lymph drainage, leading to granulomatous hypertrophy of the genitals (esthiomene), and elephantiasis. Intravaginal primary lesions will lead to intrapelvic lymphadenopathy. Subsequent fibrosis may result in rectal stricture or fistulas.

THE BACTERIA

Chlamydiae are small Gram-negative cocci. The infective particle (elementary body, EB) is 200–300 nm in diameter, and is metabolically inactive. The larger intracellular particle (reticulate body, RB) is 600–1000 nm in diameter, and is the reproductive unit. EBs are taken into the host cell cytoplasm by phagocytosis. The inclusion body moves centripetally, and the EB enlarges to form an RB. Over the next 24 hours, the RB divides to form daughter RBs, and the inclusion body enlarges to occupy most of the cytoplasm. Between 24 and 36 hours, the RBs condense down to form EBs. The mature inclusion breaks down at around 48 to 72 hours, releasing EBs into the extracellular environment. *C. trachomatis*, but not the other species, produces a glycogen matrix within the inclusion.

Chlamydiae were first isolated following the inoculation of the yolk sac of hens' eggs. This technique has been superseded by cover-slip cell culture, using chemically treated cells, e.g. cycloheximide treated McCoy epithelial cells. After 48–72 hours' incubation, cells are fixed and stained, and then examined microscopically for inclusion bodies. Chlamydiae cannot be cultured in non-living media.

Individual isolates of *C. trachomatis* are identified as serovars using specific antibodies. These are labelled A, B, C etc., and certain serovars are linked to specific clinical syndromes.

EPIDEMIOLOGY

Psittacosis is a true zoonosis. Infected birds (particularly, but not exclusively, psittacine birds) suffer a systemic disease and are characteristically listless, with signs of respiratory distress, conjunctivitis and/or diarrhoea. Animal *C. psittaci* infection is common, and is of economic importance owing to wastage of lambs and damage to bird flocks intended for human consumption.

Trachoma is widespread in some parts of the developing world, with >90% attack rates in some villages. Transmission occurs in association with poor hygiene via direct contact inoculation and by flies. *C. trachomatis* infection is the commonest bacterial genital infection in both men and women in the developed world, and the major cause of involuntary infertility. Inapparent infection is common in that it occurs in 2–4% of sexually active women attending general practice and asymptomatic cervical infection is found in 15–20% of women attending STD clinics.

LGV occurs only in the tropics, but again asymptomatic or unrecognised infection in women is the most important source of infection.

THERAPY AND PROPHYLAXIS

Chlamydiae are sensitive to tetracyclines and the macrolides (e.g. erythromycin) and these antibiotics are currently the treatment of choice. *C. trachomatis* is sensitive to sulphonamides *in vitro*, whilst *C. pneumoniae* and *C. psittaci* are not. Alternative antimicrobials include rifampicin (although resistance develops readily *in vitro*) and the fluorinated 4-quinolone ofloxacin. Chlamydiae are clinically resistant to β-lactam antibiotics, aminoglycosides and metronidazole. Prophylactic antibiotics are not indicated. Vaccination of humans has not proved successful for prevention of trachoma, and may result in hypersensitivity.

LABORATORY DIAGNOSIS

Particular care is required with specimen collection. Urethral swabs from men must be passed 3–4 cm into the urethra; the female cervix must be cleaned before sampling. The correct transport medium must be used, and specimens for culture rapidly transported to the laboratory. Scrapings are taken from the upper tarsal conjunctiva for trachoma, and swabs from the lower conjunctiva for paratrachoma.

Cell culture is currently the best method for diagnosis of *C. trachomatis* infection, but is impracticable in most settings. Antigen detection techniques are widely available, but are not ideal owing to problems with sensitivity, specificity and the predictive value of positive results, particularly in low risk populations. The most reliable antigen test available is a direct immunofluorescence test using monoclonal antibodies, but these tests are labour intensive for large numbers of samples. Less sensitive are the enzyme-linked immunosorbent assays (ELISA), using polyclonal or monoclonal antibodies.

Measurement of serum antibodies is much less satisfactory, although it is currently the most reliable method for diagnosing *C. pneumoniae* infection. Paired sera are required to demonstrate a fourfold rise in antibody titre. There is considerable cross-reactivity between chlamydial species, and great care is required when interpreting antibody levels in suspected genital infection. High specific IgM levels in neonates are diagnostic. Complement fixation tests are insensitive and genus specific, and immunofluorescence tests are preferred.

36 The Rickettsiae

The family Rickettsiaceae consists of small, aerobic Gram-negative bacilli which are obligate intracellular parasites. Members of the genera *Rickettsia* and *Coxiella* cause disease in man (typhus, spotted fever, Q fever).

TRANSMISSION All the rickettsiae are transmitted to man by arthropods. The natural reservoirs are either man or wild rodents and the vector of each specific organism is either ticks, mites, fleas or lice. *Coxiella burnetti*, the cause of Q fever, persists in domestic animals and is transmitted to man from animal placentae or a contaminated environment.

CLINICAL FEATURES The spotted fevers and typhus are characterised by fever, headache, respiratory symptoms and a generalised maculopapular rash which often becomes petechial or purpuric. In Q fever, fever and headache are again common but atypical pneumonia is the most characteristic illness.

COMPLICATIONS In spotted fevers and typhus hepatosplenomegaly, neurological complications, myocarditis and disseminated intravascular coagulopathy may occur. The mortality is 10–30%. Chronic infection with *Coxiella burnetti* may cause endocarditis.

THERAPY AND PROPHYLAXIS Antibiotic therapy with a tetracycline or chloramphenicol should be started as soon as possible; otherwise mortality is high. Vaccines are available against some of the organisms. Avoidance of the insect vectors should be practised.

LABORATORY DIAGNOSIS Serology is the mainstay of diagnosis of infection with these organisms.

CLINICAL FEATURES

Spotted Fevers

These diseases (e.g. Rocky Mountain spotted fever, Mediterranean spotted fever, rickettsial pox) have an incubation period of 7–14 days. There is a sudden onset of fever, chills and myalgia, and headache is a prominent symptom. Between 2 and 5 days later a generalised rash appears which may be maculopapular with petechiae or purpura (spotted fevers) or vesicular with pox-like progression (rickettsial pox).

Typhus

There are three forms of typhus (epidemic, endemic and scrub) with distinct epidemiology and vectors. The illnesses have incubation periods of 1–2 weeks and begin non-specifically with fever, headache and myalgia. About 5 days after the onset of illness a maculopapular rash appears in most cases of epidemic and endemic typhus but in fewer than 50% of cases of scrub typhus.

Q fever

This disease has a long incubation period of about 3 weeks and starts with fever, chills, malaise and myalgia. Headache is also prominent. Respiratory symptoms are common but usually mild although atypical pneumonia with extensive radiological changes may occur.

COMPLICATIONS

In severe cases of Rocky Mountain spotted fever splenomegaly, nervous system involvement, clotting disorders and circulatory collapse may occur. The overall mortality is about 3%. Mediterranean spotted fever and rickettsial pox are milder with recovery untreated in 2–3 weeks.

In epidemic typhus myocarditis and meningoencephalitis with delirium and coma

Table 36.1   Vectors and reservoirs of rickettsiae.

Disease	Organism	Vector	Reservoir	Distribution
Rocky Mountain spotted fever	*Rickettsia rickettsii*	Tick	Wild rodents, dogs	USA
Mediterranean spotted fever	R. *conori*	Tick	Dogs	Around Mediterranean
Rickettsial pox	R. *akari*	Mite	Wild rodents	USA, states of the former USSR, Korea
Scrub typhus	R. *tsutsugamushi*	Mite	Wild rodents	Asia
Epidemic typhus	R. *prowazekii*	Louse	Man	—
Endemic typhus	R. *typhi*	Flea	Wild rodents	Asia
Q fever	*Coxiella burnetti*	—	Cattle, sheep, goats	Worldwide

may occur and some epidemics have had mortalities of over 50%. There may be reactivation of epidemic typhus in an individual years after the initial attack. Endemic and scrub typhus are milder, generally uncomplicated diseases.

Hepatosplenomegaly may occur in Q fever with the liver containing diffuse granulomas. Subacute endocarditis may occur months or years after initial infection and is sometimes called chronic Q fever.

THE BACTERIA

Rickettsiae are small aerobic Gram-negative bacteria once thought to be viruses because of their size and the fact that they are obligate intracellular parasites. Thus in the laboratory they can only be grown in embryonated eggs or cell culture. The organisms usually die rapidly in the extracellular environment except *Coxiella burnetti*. Species which cause disease in man are maintained in animal reservoirs and are transmittd by arthropod vectors (Table 36.1).

EPIDEMIOLOGY

Man is the only known reservoir of R. *prowazekii* and epidemic typhus has historically been most widespread during wars. It is now rare but endemic and scrub typhus are relatively common in some parts of Asia. Rocky Mountain spotted fever is currently most common in central and eastern states of America.

Q fever is worldwide in distribution and the hardiness of the organism in the environment is important in the transmission of the infection.

THERAPY AND PROPHYLAXIS

Rickettsial disease should be treated as promptly as possible with tetracyclines or chlorampenicol. Mortality tends to be associated with severe disease in which treatment is commenced late.

People going to endemic areas are advised to wear protective clothing impregnated with insect repellant so as to minimise tick bites. Vaccines and prophylactic antibiotics have been experimented with but are only justified in wartime.

Acute Q fever should be treated with tetracycline. Q fever endocarditis is difficult to treat, requires combination antibiotics and may necessitate valve replacement.

LABORATORY DIAGNOSIS

The mainstay of the diagnosis of rickettsial disease and of Q fever is serology. A variety of methods are available but antibodies are relatively slow to develop. Paired sera must be tested. The Weil–Felix test detects antibodies which cross-react with antigens from *Proteus* species. It has frequently been used in the past but is insensitive and non-specific.

Rickettsiae can be isolated in cell culture or embryonated eggs but this should only be carried out by experienced personnel in appropriate containment facilities.

Index

Index compiled by Geoffrey Jones